Rural–Urban Interaction in the Developing World

Sustaining the rural and urban populations of the developing world has been identified as a key global challenge for the twenty-first century. *Rural–Urban Interaction in the Developing World* is an introduction to the relationships between rural and urban places in the developing world and shows that not all their aspects are as obvious as migration from country to city. There is now a growing realisation that rural–urban relations are far more complex.

The book takes rural–urban relations as its focus, rather than considering them as only a part of either urban development or rural development. It examines a range of interactions between the rural and the urban by considering these interactions as flows that can take place in either direction. It considers migration as just one of a series of flows between the rural and the urban, rather than only focusing on the phenomenon of rural-to-urban migration as a strong and highly visible indicator of urbanisation. Each of the flows of people, food, the environment, money and ideas has its own chapter. The book steps back from accepted orthodoxies by considering the flows as interactions that may take place in either direction, across space as well as within sectors. These flows are also considered within the context of development theory.

Rural–Urban Interaction in the Developing World uses a wealth of student-friendly features including boxed case studies, discussion questions and annotated guides to further reading to place rural–urban interactions within a broader context. It promotes a clearer understanding of the opportunities, as well as the challenges, that rural–urban interactions represent.

Kenneth Lynch is a Senior Lecturer in the School of Earth Sciences and Geography at Kingston University and holder of a National Teaching Fellowship.

Routledge Perspectives on Development

Series Editor: Tony Binns, *University of Sussex*

The Perspectives on Development series will provide an invaluable, up-to-date and refreshing approach to key development issues for academics and students working in the field of development, in disciplines such as anthropology, economics, geography, international relations, politics and sociology. The series will also be of particular interest to those working in interdisciplinary fields, such as area studies (African, Asian and Latin American Studies), development studies, rural and urban studies, travel and tourism.

Published:

David W. Drakakis-Smith
Third World Cities, Second Edition

Jennifer A. Elliott
An Introduction to Sustainable Development, Second Edition

Janet Henshall Momsen
Gender and Development

Kenneth Lynch
Rural–Urban Interaction in the Developing World

Forthcoming:

Tony Binns, Peter Illgner and Etienne Nel
Indigenous Knowledge and Development

Hazel Barrett
Health and Development

Chris Barrow
Environmental Management and Development

Alison Lewis and Martin Elliott-White
Tourism and Development

Nicola Ansell
Children and Development

John Soussan and Mathew Chadwick
Water and Development

Katie Willis
Theories of Development

Rural–Urban Interaction in the Developing World

Kenneth Lynch

LONDON AND NEW YORK

First published 2005
by Routledge
2 Park Square, Milton Park, Abingdon, Oxon OX14 4RN

Simultaneously published in the USA and Canada
by Routledge
270 Madison Ave, New York, NY 10016

Routledge is an imprint of the Taylor & Francis Group

Typeset in Times New Roman by
Florence Production Ltd, Stoodleigh, Devon
Printed and bound in Great Britain by
The Cromwell Press, Trowbridge, Wiltshire

British Library Cataloguing in Publication Data
A catalogue record of this book is available from
the British Library

Library of Congress Cataloging in Publication Data
A catalog record for this book has been requested

ISBN 0–415–25870–7 (hbk)
ISBN 0–415–25871–5 (pbk)

Contents

Figures

Tables

Boxes

Acknowledgements

Although one name is on the cover of this book as author, many people have contributed to the writing with or without knowing. Colleagues and students at Kingston University have influenced some of the ideas that have emerged in this book, through their influences on my interests, my research and my teaching over an enjoyable decade. Thanks to the various colleagues who have read or heard versions of these ideas, for their questions and comments; and thanks, in particular, to Tony Binns, as Series Editor, and Andrew Mould and Anna Somerville at Routledge. Claire Ivison, of Kingston University, has been enormously patient each time I turned up at her office with 'just one more illustration'. I am also grateful to the two reviewers who remained anonymous but whose comments helped to produce a more readable and more useful book. I also have to thank my physiotherapist, Paul Miller of Surbiton Hospital, because without him stepping in to treat the early stages of tendonitis, this book would have been much later than it was.

Thanks to Nigel Poole, Imperial College, for permission to use the photograph in Figure 2.7 and Andrew Mark Bradford, Royal Holloway, for permission to use the photographs in Figures 3.4 and 3.5. All other photographs are the author's own. Thanks to the authors, organisations and publishers who have given permission to reprint their materials in the book. The sources are indicated where appropriate throughout.

Finally, I must thank Ronan and Clare. To have had the patience to put up with a reclusive writing father and husband is bad enough, but to have this happen during the first year of Ronan's life has been difficult for them. Clare has also read every word and her comments have made for a more readable book. I dedicate the book to Ronan and his generation. They will have to continue the work that we, and previous generations, have started in tackling the world's great challenges. I hope they are able to hand on to their children a better world than the one we hand on to them.

However, in spite of numerous helpers and supporters, at the end of the day the ideas and the mistakes in this book are my own responsibility.

Surbiton
May 2004

Introduction

This book sets out to reunite the urban and the rural areas in the study of development across the developing world. Most approaches in development studies – both theoretical and empirical – are based on the premise that there is a clear distinction between the urban and the rural. However, this distinction has been challenged. There is research on 'rural' activities in 'urban' spaces, urban activities in rural spaces, and on the changing interface between urban and rural spaces and on the increasing interdependence between these two realms. There is therefore a need to bring these disparate themes together in one volume.

Some of the earliest works on the interaction between city and country in the developing world focused on modernisation diffusion (Gould, 1969; Rostow, 1960). These were developed into spatial models that may be pessimistic, for example Friedman's (1966) core–periphery model, or optimistic, for example Vance's (1970) mercantile model. These influential theories are primarily focused on settlement hierarchies rather than on the interaction between town and country, suggesting an urban focus, although they are used to theorise about rural–urban interaction. Although not the originator of this concept, Lipton (1977) made a considerable impact on development studies later, presenting a thesis on the ways in which urban-based industrialisation policies can have an adverse impact on the development of rural areas. Subsequently, a number of studies looked into Lipton's ideas of urban bias, some agreeing that urban bias undermined rural development, some arguing that the distinction between urban and rural areas was rather crude and did not reflect the complex reality. More recently, however, the theory of the distinction between urban and rural development has been questioned in the context of a number of disparate themes. This questioning has been particularly strong in the field of demography and migration studies. The impact of the economic crises of the 1980s has also prompted research on the differentials between cities and rural areas. Continued urban growth has prompted concerns about the environmental impacts on the countryside. In consequence regional development initiatives have been based on small towns, and analyses have emphasised the importance of maximising rural–urban interaction for development. Disparate critiques have begun to coalesce into a stronger body of research in recent years.

This book does not necessarily break new ground but serves to collate a wide range of research and theory which has relevance for this question of interactions between urban and rural areas. The book makes much of this material more easily accessible to a wider audience; in particular it is aimed at second- and third-level undergraduate students of geography, development studies, sociology, economics and planning. It is also of relevance to postgraduate students.

The current emphasis of research appears to be moving away from the study of the physical interface between urban and rural, in favour of a stronger focus on the relationships between them. This is even the case in recent approaches to the peri-urban interface where rural and urban come into contact (Tacoli, 1998a). Even in the physical space where the urban and the rural meet there is an emerging consensus that the physical location of these linkages is less important than the way they are constructed and structured (see the discussion of the importance of institutions to an understanding of natural resource management in the peri-urban interface in Chapter 3). This book therefore focuses on flows between urban and rural areas. This encourages a consideration of the movement of goods, people and ideas across the interface between cities and the countryside.

This theme of flows and linkages recurs throughout the book. In spite of the challenge of the fluidity and fragmented identities that play a role in the lives of the people who communicate, exchange and travel across the urban–rural divide, a key contention is that this fluidity is often a deliberate strategy of those living in rural and urban areas in order to maximise their livelihood opportunities. This focus on flows and linkages is part of the philosophy that has been adopted in the structuring of this book. Chapter 1 introduces the approach adopted by this book, explaining how it will focus on the flows that link the areas rather than the structures or processes that separate them, as this provides a more useful and more powerful analysis of the relationship between city and countryside.

The remaining chapters therefore each focus on an aspect of these flows. Chapter 2 focuses on flows of food, specifically on the supply of food to the city from the countryside, as well as on the interaction with urban-based food procurement strategies which may include urban as well as rural production. Chapter 3 focuses on systems of the natural environment and resources, including the systems that provide cities with their raw materials and energy. This chapter also examines the implications of obtaining these resources from rural areas and the extent of their ecological impact, raising the question of how far a city places its environmental burden beyond its own boundary. Chapter 4 examines human strategies and how these impact upon people's residence and movement from urban to rural or rural to urban areas. Chapter 5 examines the importance of ideas and information. Finally, wealth flows are discussed in Chapter 6.

As most studies of development tend to take place in either urban or rural locations, there is much in this book that is relevant to students of both areas. There is also much that relates to the topic of this book which is published elsewhere in the Routledge Perspectives on Development series. This book therefore contains many references to these other texts, in particular where they provide appropriate examples.

1 Understanding the rural–urban interface

Summary

- **Past approaches to development studies have tended to focus on either urban or rural spaces.**
- **New development paradigms consider networks and flows, so it is important to reconsider flows and linkages between rural and urban areas.**
- **Some rural–urban links can favour one area or the other, but it is important to be aware that the net benefits can flow both ways, resulting in change both over time and from one place to another.**
- **Urban–rural links have been important to development theory although this topic is rarely a focus of development research.**

Introduction

This chapter introduces key ideas that will be the building blocks of the later chapters in the book. It sets out to explain why it is important to understand urban–rural relations in the developing world, how they relate to the broader evolution of development theory and how such study might help us understand the problems of development and poverty in some of the world's poorest countries. The chapter concludes with a brief explanation of the rationale of the way the subject has been divided into the chapters in this book.

Why it is important to study rural–urban interaction

The rapid population growth of Third World cities gives rise to concerns about the changing nature of the relationship between urban and rural. The evidence for this is in a growing number of recent publications, research reports and policy documents of international organisations, which emphasise key development concepts such as decentralisation (see also Table 1.1 and Box 1.1). The UNCHS (1999) estimated that the world's urban population would be 2.9 billion in 2000, accounting for 47 per cent of the global total. This is an increase on the 30 per cent in 1950; the urban population is likely to go over 50 per cent in 2007.

Table 1.1 ***Summary of international agency initiatives on rural–urban linkages***

Organisation	*Activity*	*Weblinks for details*
World Bank Economics Division	Research papers	– Lnweb18.worldbank.org/essd/essd.nsf/rural development/portal – http://www.worldbank.org/urban/urbanrural seminar/
UNFPA	Chapter in *State of World Population* 1996	– http://www.unfpa.org/swp/1996/ch5.htm
DFID Peri-Urban Interface	Research projects in Kumasi, Ghana, and Hubli-Dharwad, India	– http://www.ucl.ac.uk/dpu/pui/
IIED	Urban–rural linkages research	– http://www.iied.org/rural_urban/index.html – http://www.unhabitat.org/HD/hdv5n1/contents.htm
UNCHS	Promoting rural–urban linkages	– http://habitat.unchs.org/home.htm
OECD Club du Sahel	Research on regional integration and the development of local economies (ECOLOC)	– http://www1.oecd.org/sah/activities/Dvpt-Local/DLR9.htm
FAO's Food into Cities Programme	Food links research and policy recommendations	– http://www.fao.org/ag/ags/agsm/SADA/SADAE-5_.HTM

Note: DFID = Department for International Development. IIED = International Institute for Environment and Development.

Source: adapted from World Bank (2002).

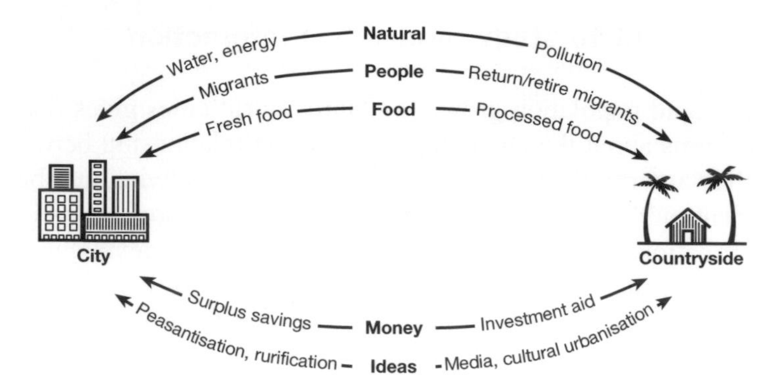

Figure 1.1 ***Rural–urban interactions.***

Figure 1.1 illustrates schematically the way rural–urban flows have been organised in this book. However, the figure has limitations, as most two-dimensional representations of complex systems have. One of the key ideas the figure presents is the possibility of each of the flows to work in either direction. Under certain conditions one-way flows may dominate, or the emphasis may change over time or from one context to the next. The flows of people and food and natural flows are represented above ground to acknowledge their visibility. Money and ideas are less tangible and involve service sectors in delivery, so are represented below ground as they are less obvious. One of the main limitations of the figure is that it represents city and countryside as clearly separate. This ignores more problematic issues such as definitions of the rural–urban interface, where distinctions between city and countryside can become blurred. Another important theme that is discussed in this book is the role of intermediate settlements, which are not represented in the figure. Finally, examples of the types of flows provided in the arrows are selective. More detailed examples are discussed later in the book.

Until relatively recently much rural–urban research has focused on a single city and its hinterland. However, the increasing importance of international links and the process of globalisation have an impact on rural–urban relations. In addition, cities throughout the world are often caught between the pressure to be included in the world economy, on the one hand, and on the other the need for links with their rural hinterlands. Such tensions raise issues that have not been considered until recently. This tension is paralleled in the competing processes of globalisation and decentralisation.

A further reason for writing a book that focuses on the relations between urban and rural areas relates to the fact that these relations have been important throughout history. Many major world developments have been linked to the relations between urban and rural realms. For example, on coming to power in 1949, the revolutionary Chinese government was faced with one of the world's greatest urban–rural dilemmas. It was made up of politicians largely from peasant backgrounds and with peasant support, but it took control of a country already struggling to cope with the demands of its many large cities. By 1950 Shanghai was already estimated to number 5.3 million people, Beijing 3.9 million, and Tianjin 2.4 million (UNCHS, 2001). Some of the earliest government actions were designed to impose state-controlled marketing of agricultural goods and rationing of urban food consumption, and to control the movement of people – especially rural–urban migration. In addition, the level of land scarcity experienced by rural dwellers convinced the Chinese government that urban-based heavy industrialisation was the only way the country would be able to support its population which was 541 million in 1952, of whom 57.7 million, or 10.6 per cent, were urban (Knight and Song, 2000). One of the issues that worried the Chinese government was that in 1964, the second census under communist rule found that the urban population had grown to 129.3 million, or 18.4 per cent. Such rapid growth, both nationally and in the cities, alarmed the government, which decided that there was a need to control population growth and to create employment for the urban population.

The hostility of external global powers meant that foreign investment was unlikely. Therefore the only possibility of producing savings to invest in industrialisation was to extract surplus from rural agricultural production, resulting in a bias of policy towards the urban rather than the rural areas. The result was that the State Purchase and Marketing Cooperative's main function was to 'extract as much of the harvest as possible from the peasants' (Knight and Song, 2000: 11). The Cooperative also supplied agricultural inputs and the state controlled the banking. The state therefore mediated all rural–urban flows of goods and capital other than household remittances. The pro-urban bias was compounded by the agricultural tax which amounted to approximately 30 per cent of farm proceeds (Knight and Song, 2000). The Chinese government therefore funded its industrialisation by mobilising the rural areas to produce and save more in order to provide the capital and the tax revenue for investment.

Box 1.1

Comparing international organisations' approaches to rural–urban linkages

A number of national and international aid agencies are beginning to revisit the issue of rural–urban interaction. There appears to be considerable convergence in the thinking and approach at this level. For example, much of the intervention in which the World Bank engages is guided by separate rural and urban strategy documents. However, a recent workshop arrived at the following overarching themes (World Bank, 2000):

- the need for a broad analytical framework that can integrate the processes and approaches that span the realms at various scales;
- the need to consider the role that spatial dimensions play, along with the dynamics, vulnerabilities and movements in and out of poverty;
- the importance of a long-term perspective in relation to shifts in settlement and economic patterns;
- recognition that local economies advance and decline and that different approaches may therefore be required, and the need to understand the linkages and their role in the changes;
- the need to recognise the importance of the trend towards decentralisation;
- the need for approaches that facilitate working across sectors;
- the need to recognise the heterogeneity of constructs of town and country and the divisions between them, including agriculture in towns and non-farm activities in the country.

The UK's governmental aid organisation, the Department for International Development, has supported some research in this area. A recent summary sheet outlined the major challenges for rural–urban linkages as:

> - The poorest areas may have little more than consumption linkages
> - Production linkages emerge in more diversified settings, such as where rural-based workshops start to supply urban-based factories
> - Financial linkages appear in all settings, but with different outcomes for rural economies
> - The rise of network societies may contribute to bypass effects, when financial flows link rural areas directly with distant, larger cities at the expense of local towns.
>
> (ODI, 2002)

Common ground is emerging here. There is a consensus that more flexible approaches are required, that flows may wax and wane over time, that the role of small towns in rural–urban interactions must be considered and that interventions in either rural or urban areas have implications beyond their locality. One of the challenges appears to be that it is not always clear which department or division should be responsible for the agency approach, with some agencies considering an issue from a rural-focused division and others from an urban-focused division.

In a recent report Mutizwa-Mangiza (1999), Planning and Coordination Officer for UNCHS, highlights past views on the rural–urban divide as either pro-urban or anti-urban. According to Mutizwa-Mangiza, these biases have underpinned much development policy and intervention affecting both urban and rural areas. However, he argues that promoting rural–urban linkages offers considerable potential for developing the entire rural–urban continuum. He highlights the importance of this balanced approach, referring to its place in the Habitat Agenda agreed at the Habitat II Conference organised by UNCHS in Istanbul in 1996.

- (a) 'To promote the sustainable development of rural settlements and to reduce rural-to-urban migration . . .' (para.165);
- (b) 'To promote the utilization of new and improved technologies and appropriate traditional practices in rural settlements development . . .' (para.166);
- (c) To establish '. . . policies for sustainable regional development and management . . .' (para. 167);
- (d) 'To strengthen sustainable development and employment opportunities in impoverished rural areas . . .' (para.168); and
- (e) To adopt 'an integrated approach to promote balanced and mutually supportive urban–rural development . . .' (para.169).

(Habitat, 1996, quoted in Mutizwa-Mangiza, 1999: 6)

According to Mutizwa-Mangiza, achieving these objectives will require:

i) strengthening of rural–urban linkages mainly through the improvement of marketing, transportation and communication facilities;
ii) improvement of a number of infrastructure components which, while enhancing rural–urban linkages, are also essential for economic growth and employment creation (both farm and non-farm) within small urban settlements and rural areas themselves, especially roads, electricity and water;
iii) bringing private and public services normally associated with cities to the rural population; and
iv) strengthening of sub-national governance at the regional, rural-local, and city-region levels.

(Mutizwa-Mangiza, 1999)

While the Chinese example illustrates a deliberate policy of exploiting rural areas, Hodder (2000) argues that inevitably rural agricultural sectors and urban industrial sectors play strategic roles in each other's development. He identifies six key reasons for these close links between the two sectors (2000: 80–82):

1. Agriculture depends on manufactured goods both for the transformation of produce (for example, farm tools, machinery, inputs) and for the consumer goods which are in demand as agricultural incomes rise (such as radios and bicycles).
2. As agriculture incorporates more technology in its activities, labour becomes a less significant factor. More technologically advanced agriculture releases capital and labour which move into the urban industrial sector.
3. Agriculture provides raw materials for some industries, such as tobacco, cotton and sisal.
4. Agriculture for export can earn foreign exchange which is important for purchasing items which are vital to industrial processes. These include commodities such as petroleum, chemicals and technology which is not produced locally.
5. There is an important balance to be struck in incomes, prices and taxation between the urban and the rural areas. For example, high food prices provide rewards to farmers and incentives to increase production, but may mean high prices in urban areas which can lead to poverty and unrest. Taxation in the agricultural sector may be necessary to raise revenues to finance public expenditure, but may act as a disincentive to farmers, particularly if much of the expenditure is urban or industrial focused.
6. In rapidly urbanising countries agriculture produces strategically important food for the growing number of urban residents, thus ensuring food security at prices that are affordable.

International and national agencies are beginning to realise that their programmes of urban and rural development intervention have impacts on each other and that there is a need to integrate some of the thinking behind them (see Box 1.1).

The six reasons listed above explain the key interdependencies between rural and urban areas. While the discussion has so far concentrated on the flows between the urban and the rural, one of the main reasons for the separate approaches is the attempt frequently made to identify the defining characteristics of what is 'urban' and what is 'rural'. However, Rigg (1998a) cautions against 'pigeonholing' when it comes to such definitions. For example, he argues that to separate people into urban or rural categories is problematic. He outlines three main difficulties associated with such categorisation.

1 Registration records often do not detect changes in residence. For various reasons it may be undesirable for recent in-migrants to be registered as urban dwellers. Rigg gives examples of under-reporting of urban residence, particularly in relation to the controversies this can pose during elections when how an area is defined can have implications for the number of political representatives or the authority into which they are elected (for further discussion of this see Chapter 4 below).
2 Allocating people to discrete categories such as 'urban' or 'rural' assumes that these categories accurately reflect their realities. Rigg's own empirical research in Thailand (1998b), among others, has demonstrated the significance of fluid, fragmented and multi-location households to survival strategies. This results in households straddling and moving across the rural–urban interface. Thus categorising them as one or the other makes no sense. In addition, there is the problem of defining the boundaries of urban places which is usually done on some arbitrary basis. This will be discussed in more detail in Chapter 4 below.
3 Rigg argues that many Asian urban residents do not consider the cities and towns they live in as 'home'. This is because they ultimately intend to return to their rural origins. This, he argues, brings the issue of the identities of individuals into focus and '"home" and "place" are ambiguous and shifting notions, where multiple identities – both – can be simultaneously embodied' (Rigg, 1998b: 501).

A final concern that could be added to Rigg's three points above relates to the blurring of the actual geographical definition of the rural–urban divide. This is particularly the case where cities are expanding rapidly and extending their physical limits and influence outwards into the rural areas. To add to the difficulty of definition, we find that different sizes of settlement are defined as 'urban' by different countries. For example, in China, relatively small settlements are not counted as urban (see Box 1.2). However, a city, or city-region, may include a conglomeration of areas extending beyond the city proper. While such data make it possible to monitor demographic changes in countries across time – for example,

Table 1.2 ***Rural population densities of selected countries***

Country	*Rural population (% of total)*		*Average % annual growth*	*Rural population density (people per sq. km of arable land)*	*Land area (1,000 sq. km)*	*Arable land (% total land area)*		*Permanent cropland (% of land area)*	
	1980	*2000*	*1980–2000*	*1999*	*1999*	*1980*	*2000*	*1980*	*2000*
Bangladesh	86	76	1.5	1,209	130	68.3	62.2	2.0	2.6
Burundi	96	91	2.2	792	26	35.8	30.0	10.1	12.9
China	80	68	0.4	691	9,327	10.4	13.3	0.4	1.2
Colombia	36	25	0.2	508	1,039	3.6	2.0	1.4	2.2
Egypt	56	55	2.1	1,217	995	2.3	2.8	0.2	0.5
Indonesia	78	59	0.4	694	1,812	9.9	9.9	4.4	7.2
Rwanda	95	94	2.4	901	25	30.8	35.1	10.3	10.1
Sierra Leone	76	63	1.3	653	72	6.3	6.8	0.7	0.8
Sri Lanka	78	76	1.2	1,660	65	13.2	13.6	15.9	15.8
Tanzania	85	72	2.1	640	884	3.5	4.2	1.0	1.0
Vietnam	81	76	1.6	1,031	325	18.2	17.7	1.9	4.9
Low income	76	68	1.6	510	32,536	11.8	13.2	1.0	1.4
Lower middle income	69	58	0.5	642	43,596	8.8	9.2	1.0	0.9
Upper middle income	38	24	–0.6	184	23,048	7.0	8.0	1.1	1.3
High income	25	21	–0.1	180	30,920	12.0	11.6	0.5	0.5

Source: adapted from World Bank (2002).

comparing rates of urban growth from one census to the next – this problem of differences of definition suggests that comparisons between countries are problematic to say the least (see Table 1.2). For example, Afsar (1999) reports that Bangladesh, the world's ninth largest country by population, experienced urban population growth of 6 per cent per annum between 1970 and 1996 and the proportion of the population living in urban areas grew from 7.4 per cent to 20 per cent of the total. However, she points out that the designation of new urban areas accounted for 8 per cent of urban growth from 1961 to 1974 and this increased to about one-third between 1974 and 1981. The result is that the progression from rural to urban or vice versa is very unclear and varies from one country to another. The fact that such enormous populations as the Chinese and Bangladeshi are reallocated arbitrarily from rural to urban areas raises questions about the validity of globally aggregated data on urban growth rates and urbanisation. Hardoy *et al.* (2001) suggest that it is best to consider the proportion of urban and rural populations not in terms of precise percentages because of the difficulty of enumeration, but rather in terms of a range.

This concern over the headline figures is reinforced when two more issues are considered. First, static definitions of 'urban' and 'rural' fail to take account of how the seasonally migrant populations, who move between urban and rural in some countries, are classified. Second, it is important to consider the definition of the physical boundaries between the urban and rural areas and the problem which prompted the development of the concept of the 'urban fringe' and the 'peri-urban interface'. Should these be classified as urban or rural? These issues are not consistently defined around the world and so it is in fact problematic to consider such data as comparative. There is a range of approaches to defining them (see Table 1.5). Beyond the issue of numbers is the fact that assumptions are made about people's livelihoods and activities in terms of whether they are 'urban' or 'rural', when in reality the key issue for the people concerned is the opportunity for livelihoods. These ambiguities are to an extent accepted by major international donor agencies, as illustrated in recent documents and discussions (see Box 1.1).

It is therefore vital to consider the following discussions in the light of these four concerns. One approach that is put forward by this book is to focus not so much on the boundaries between rural and urban areas as on the links between them. This serves to emphasise the interdependence of town and country. It will allow for the assessment

Box 1.2

Urban definitions in China

Most authors writing about either population or urban issues in developing countries draw attention to the problems of population data, which in many cases are a result of the general problems of data collection. However, one challenge that is often overlooked, especially when considering aggregate population data at a world or regional level, is the variation in definitions of urban and rural areas, a variation not only between countries but over time. There is more discussion of this in Chapter 2 below. In China, a number of changes in the definition of urban areas have taken place through time which affect the available data (see Table 1.3). This is particularly important when considering the global situation because China accounts for more than one-sixth of the world's population. Changes to the definition of what areas are rural and what are urban can have an effect on the headline categorisations of the urban and rural populations of the country.

The result is that the urban population growth as a result of definition change is considerable. Rakodi (2002) estimated that 40 per cent of China's urban population growth was due to changes in the urban administrative system. This is illustrated in Table 1.4, which shows how significant such definition changes can be. For example, the 1982 definition resulted in a drop in the rural population from 932 million to 535 million. This raises questions about changes in other aspects of the development process in China, for example the country's much-discussed 'rural' industrialisation initiatives and

Table 1.3 ***China's changing definitions of an urban area***

Prior to 1964	An area of more than 2,000 permanent residents of whom more than 50% are non-agricultural
1964 revision	(1) An area with more than 3,000 permanent residents of whom 70% or more are non-agricultural (2) An area of more than 2,500 of whom 85% or more are non-agricultural
1982 revision	(1) An area of county-level government agency (2) A township totalling less than 20,000, but where the non-agricultural population is 2,000 or more (3) A township of more than 20,000 where the non-agricultural population is greater than 10% (4) A remote, mountainous, mining, harbour or tourism area where the non-agricultural population is 2,000 or more
1990 revision	(1) All residents of urban districts in provincial and prefectural-level cities (2) Resident population of 'streets' (*jiadao*) in county-level cities (3) Population of all residents' committees in towns.

Source: adapted from Heilig (1999).

the extent to which 'urban agriculture' would be defined as urban in another country using different definitions of 'urban' and 'rural'.

Table 1.4 ***Rural population in China by selected definitions***

	Total population	*Pre-1982 definition*		*1982 census definition*		*1990 census definition*	
	(million)	*(million)*	*(%)*	*(million)*	*(%)*	*(million)*	*(%)*
1978	962.59	838	87.1	790	82.1	n.a.	n.a.
1982	1,016.54	869	85.5	805	79.2	n.a.	n.a.
1987	1,093.00	895	81.9	584	53.4	n.a.	n.a.
1990	1,143.33	932	81.5	535	46.8	842	73.6

Source: data from Heilig (1999).

Table 1.5 ***Examples of prevailing definitions of the peri-urban concept***

Spatial/Locational	• Based on *distance* from the city centre and relative to the built environment, e.g. peri-urban as those zones at the edge of the built-up areas • Draws on *land use values* and proportion of non-agricultural activities in the land uses • Considers an area or activity in terms of the legal or administrative boundary of the city, those just outside being peri-urban
Temporal	Areas recently incorporated into the city or that are contiguous to the city and whose use (usually built development) is recent or below a certain age (maybe 5-10 years).
Functional	Areas that may be outside the city boundary but are functionally integrated or linked to the city on the basis of certain criteria and cut-off points, e.g supply of fresh produce to the city, daily commuting to the city, labour participation, etc.
Social exclusion	A definition also based on linkages but looking at areas and social groups within the city. The peri-urban are those areas and social groups located within the city boundary but are socially, economically and functionally excluded from the rest of the city. Criteria could be: • *Infrastructure*: such exclusion is usually assessed on the basis of infrastructure provision (water and sanitation being the *most* common) • *Informal settlements* are also used as an indicator of exclusion
Conflict	A view that is analytical and considers peri-urban areas as places of conflict where two or more different systems clash, as opposed to the convergence and harmonisation of different systems: • rural vs. urban • agriculture vs. built development • modern vs. subsistence • formal vs. informal

Source: adapted from Mbiba (2001).

of relations that may favour one realm or the other (Harriss and Harriss, 1988; Lipton, 1977, 1984, 1993; Pugh, 1996). It will also allow for some ambiguity about the boundaries between the realms and about the identities of those who live in, between or across the so-called rural–urban 'divide' (Harriss and Moore, 1984; Tacoli, 1998b) or 'gap' (Jamal and Weeks, 1988). This approach will provide more powerful and more flexible understandings of the geographies of the relations between the urban and the rural, as it focuses more on the relations and overlaps than on the divides. Indeed, the growing interest in the fringe between the urban and the rural suggests there is no divide, but an overlap which provides both positive and negative flows in either direction, sometimes referred to as 'peri-urban' (Briggs and Mwamfupe, 2000).

Researchers have also presented arguments for the separate definition of high-density rural areas that can occur under a particular set of characteristics. For example McGee (1991) describes a kind of high-density rural area which he calls *desakota*, arguing that it is peculiar to Asia and suggesting that it forms as a result of the metropolitan urban area extending its economy and influence into the surrounding rural areas and creating an intense mixture of agricultural and non-agricultural activities (see later discussion). Examples include areas around Jakarta, Manila and Bangkok. Mortimore (1989, 1998) carried out research on the environmental implications of what he called the 'close settled zone' around Kano in northern Nigeria, a densely settled area that is extensive but has maintained its rural nature. Qadeer (2000) argues that it is important to consider such areas because some countries have rural population densities which are higher than 400 persons per square kilometre, a threshold used by some countries as a minimum for defining an area as urban. Table 1.3 above provides an illustration of how China has changed its definition of 'urban'. Qadeer (2000) examines examples of 'ruralopolises' comprising most of Bangladesh, from West Bengal to Uttar Pradesh and Kerala provinces in India, and the Punjab and Peshawar provinces of Pakistan; other examples he mentions include the South Yangtze Valley in China, the Mekong Delta in Vietnam, and the lower Nile Valley in Egypt. He concludes that a ruralopolis is an alternative route to urbanisation, bringing about similar social transformations and spatial reorganisation in a rural setting, and 'the future form of human habitat in large parts of Asia and Africa in the 21st Century' (Qadeer, 2000: 1601).

There is evidence that suggests that cities in the developing world rely more heavily on their own hinterlands than do cities in the developed world (Snrech, 1996; Guyer, 1987), particularly as the majority of the national population is still rural, so their relationship with their hinterland is very important. This link is reciprocal, since the rural primary production that makes up much of the export production of most developing countries usually has to be distributed through the cities. Meanwhile, the cities represent important sources of processed food, finance, services, information and imported products such as manufactured goods and some agricultural inputs. Cities therefore represent the links between rural areas and world markets. These interdependent links can form the basis of positive cycles of interaction (see Box 1.3).

A recent development brought about by colonialism which has added a further layer of complexity to rural–urban linkages is that cities in the developing world often form the interface between the rest of the country and international markets. This confuses the relationship between the city and the country. For example, Potter *et al.* (2004) discuss models of colonial space economies which illustrate how the territory is organised to suit the export of raw materials to the colonial power.

Much national infrastructure was developed to support these rural–urban linkages, making many of the cities a focal point for export production. The legacy of this is that many developing countries are left with a structure that places one city, or a small number of cities, at the heart of the nation's activities. Overlapping relationships between city and countryside are therefore a legacy of an exploitative pattern which, some have suggested, has reproduced a neo-colonial relationship between town and country (see for example Slater, 1974). Changes brought about during the colonial period, when many towns and cities were essentially internally focused (Gilbert and Gugler, 1992), meant that external concerns were forced upon them as the colonies' economies were adapted to suit the needs of the metropolitan cores. This changed during the middle of the last century when 'import substitution' and later 'export production' policies were introduced by newly independent countries in order to develop their own industrial base. Examples of countries pursuing such policies of import substitution include Brazil, whose importation of capital goods fell from 65 per cent in 1949 to 10 per cent by 1964, while its importation of consumer durables declined from 65 per cent in 1949 to 2 per cent in 1964 (Simpson, 1996).

Then during the 1970s the Brazilian government pursued a policy of export production raising its manufactured exports from 3 per cent in 1960 to 26 per cent in 1977 and 56 per cent by 1991 (Simpson, 1996).

In the case of China, discussed earlier, during the Great Leap Forward the communist government promoted industrialisation of both the cities and the countryside, raising the industrialisation and industrial employment of China as a whole. However, the massive government investment amounted to a subsidy and most industries were running at a loss by the time of Mao Zedong's death in 1976. Since then Communist Party leaders have opened China to inward investment and the industrial sector has grown very rapidly, at an average of 14.4 per cent, with exports growing at 11 per cent during the 1980s (Simpson, 1996). In both Brazil and China, during the period of import substitution an industrial base was developed which later turned into an export-producing sector.

Such policies, though intended to engender national development, focused again on the city (see Riddell, 1997). Whereas China's industrialisation had been focused on maintaining people on the land, the later export production policy has focused far more on the cities. By 2000 Shanghai had grown to 12.9 million, Beijing to 10.8 million and Tianjin to 9.2 million (UNCHS, 2001). To some extent this focus on the dominant urban core – often a primate city – resulted in the neglect of the smaller towns and prompted responses which attempted to move the focus from the dominant urban areas. In the case of Brazil this has resulted in a strongly regional concentration of growth in the south-east around the huge cities of Rio de Janeiro and São Paolo, leaving much of the rest of the country experiencing such poverty that Brazil is often counted as one of the most unequal economies in the world. The case of Brazil formed the basis for much of Frank's (1967) underdevelopment theory which argues that the development of the core areas in the south-east led to the active underdevelopment of the rest of the country. However, such industry-focused policy responses were based on research that saw an urban focus as ultimately counter-productive, apart from the fact that it favours city dwellers over rural inhabitants. Hinderdink and Titus (1998) argue that whether your perspective is pro- or anti-urban, the role of small towns is a very important and much neglected element in rural–urban interactions. In particular Rondinelli (1983) emphasised the developmental role of secondary cities in providing a range of services for their hinterlands. This prompted the so-called 'growth-poles' policies which developed intermediate settlements as poles around which growth would be encouraged to balance the dominance of the urban core. Such an argument suggests that interactions can focus on small and intermediate towns in order to encourage more beneficial relations between core and periphery. This kind of approach, focusing on the potentially positive interactions between cities and countryside, is presented by the Club du Sahel as absolutely essential to the future of West Africa because of its rapidly growing urban population. This issue is illustrated further in Box 1.3.

One factor that is particularly important for an understanding of developing areas is that urban and rural dwellers have very different ways of life. This makes comparative analysis very difficult. Urban dwellers tend to have a more monetary-based existence in which consumption practices are limited by income and decisions are defined more comprehensively by income in relation to the cost of living. In rural areas, by contrast, there is evidence that while income is important, rural dwellers are able to juggle a range of livelihood strategies based on their economic, social, cultural and ecological capital (Bryceson, 1999). This situation is further complicated by the growth in non-farm income in rural areas (Bryceson, 1999) and the 'peasantisation' of urban life (Roberts, 1995). For example, the practice of urban agriculture is reportedly widespread (Lynch, 2002), the growth in informal-sector activities is a major part of urban economies and the number of households which deliberately aim to straddle the rural–urban divide in an attempt to diversify and maximise their opportunities for sustaining their livelihoods appears to have increased (Potts, 1997). In part this may be explained by the fact that rapid urbanisation processes in developing countries mean that within the last 30 years up to half the urban population may be rural in-migrants.

Understanding development

Potter *et al.* (2004) argue that geographies of development are about relationships between people, environment and places in different locations and at a variety of scales. The nature of the interaction between city and countryside is at the heart of this focus. The relationships, by implication, involve geographies of groups and individuals. The location of a household, whether city or countryside,

Box 1.3

Urbanisation in West Africa

The Club du Sahel's programme of research known as Evaluation et Prospective de l'Economie Locale (ECOLOC) shows the impact of rapidly growing cities in West Africa. Their early report, the *West Africa Long Term Perspective Survey* (Club du Sahel, 1994), found that growth in urban areas was going to be a major challenge to the future

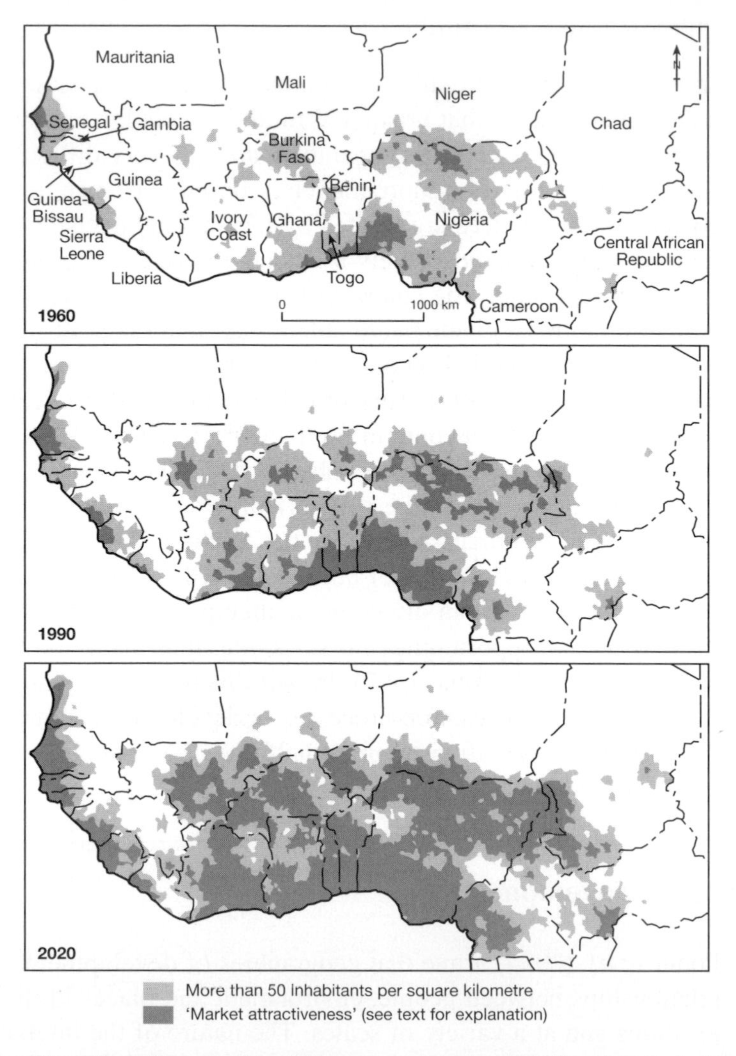

Figure 1.2 ***The interaction of urban influence and rural population density in West Africa.***

Source: Snrech, S. *Etats des réflexions sur les transformations de l'agriculture dans le Sahel.* OECD/Club du Sahel 1996. Reproduced by permission of the OECD.

of the region, the total population of which grew from 85 million to 193 million between 1960 and 1990, with the number of settlements of more than 100,000 inhabitants rising from 17 in 1960 to 90 by 1990. The report projected that at current and predicted rates there will be 300 cities of over 100,000 in the region by 2020, resulting in the merger and overlap of city hinterlands. Figure 1.2 illustrates the increasing influence of the urban over the rural areas of West Africa in 1960 and in 1990, and projected forward to 2020. There are two shaded areas represented in the maps. The darker area is what the Club du Sahel researchers describe as 'market attractiveness', or an indicator of market density, and is based on comparisons of price data. The lighter shaded area indicates a rural population density of 50 persons per square kilometre or more (Snrech, 1996). Their conclusions were that the positive impact of market density as a result of proximity to urban areas is greater than the negative impact of the increased rural population density. The implications of this for the rural areas are important.

First, the report argues that as the cities expand, so the rural areas close to them begin to find that they benefit from better market information and lower transaction costs, but are affected by higher labour costs because of non-farm alternatives and greater pressure for land because of higher population density. The report argues that the balance of evidence suggests that the benefits outweigh the disadvantages. The report goes on to state that there is evidence that some of the negative impacts are the result of time-lags in responding to the growing and changing market situation. However, this time-lag peaked during the 1980s with bad weather, high commodity export prices and ineffective national food policies. This pattern is matched by food imports as a percentage of the total food availability which has shrunk considerably since the 1980s, never havng reached more than 15 per cent. As a consequence, the report concludes that: 'it is only when they have reliable access to food markets, where outlets are sufficiently regular and profitable, that rural dwellers develop real strategies of commercial farming of food surpluses' (Snrech, 1996).

This study therefore provides evidence of the importance of the interdependence of urban and rural areas. Their prognosis for the region is that there can be a positive outcome for the rapid population growth and urbanisation. This is mainly a function of the evolution of large urban markets that will result in dynamic rural agricultural production areas. However, plans have to be put in place to ensure that the rural areas are prepared to take advantage of these opportunities. They should include the dissemination of strategies for the management of soil fertility, the intensification of agricultural production, and the organisation of farm labour; finally, the requirements for each region will be different, so these strategies must be flexible. The limitations of large parts of the region for agricultural production should not be ignored. However, they have resulted in a relatively mobile population seeking out areas of relatively high agricultural potential. Areas of population density of more than 50 inhabitants per square kilometre account for only 7 per cent of the region's territory, but almost 40 per cent of the population (Snrech, 1996). In other words, according to this work, the key to the region's future is a dynamic and flexible set of positive interactions between the urban and rural areas.

has a bearing on its livelihood assets and opportunities. This is discussed in more detail later. The interaction between urban and rural is initiated in an attempt to take advantage of the differentials or complementarities between the two areas. The differentials may be in the form of income and cost of living, supply and demand or security and hazard. Potter *et al.* (2004) go on to argue that the 'nature and relative significance of these relationships are changing constantly, both through time and space, and are themselves determined to a large extent by complex movements and flows of people, commodities, finance, ideas and information' (Potter *et al.*, 2004: 319). This book sets out to examine such complex flows and relations, focusing on the links involving urban and rural areas.

The relationship between rural and urban areas has been an important part of underlying assumptions of development theory and practice. The nature of the relationship is interpreted differently according to different theoretical perspectives. This can be illustrated by contrasting two frameworks used for analysing colonial and post-colonial development. The first is put forward by Gould (1969) who saw development as a process of modernisation. At the time, he was writing on applied quantitative methods which were the focus of an emergent approach in geographical research. Gould, being in the vanguard of this approach, applied these methods to the analysis of modernisation diffusion in Tanzania. He divided the country into a series of cells of equal area and recorded indicators of modernisation, such as miles of metalled road, number of government offices, and number of hospital beds. He did this for a selection of time periods throughout the first half of the twentieth century and mapped the changes through time and space. Not surprisingly, he found a strong bias towards the cities, in particular Dar es Salaam, then the capital. As the century progressed only the most remote areas beyond the main feeder roads remained untouched by modernisation. There was evidence, according to Gould, of a diffusion of modernisation from the urban core of Dar es Salaam down the urban hierarchy and then out into the rural areas. Arguably this is evidence of what has been described elsewhere as the 'trickle-down' effect.

However, Slater (1974) approaches the same period and the same colony and then country in a more radical way. He argues that rather than the country experiencing modernisation, what is in fact occurring is a far more sinister process that suits not the Tanzanians in the rural areas, but those in the metropolitan core, in this case the elites in Dar es Salaam and the colonial power, Britain. According to Slater, therefore, three key processes occur during the period of colonisation. First, the colonialists engage in a process of *penetration*, as they penetrate the territory and adapt it to suit their home markets. As the colony develops, so the need for labour and resources becomes greater and there is a realisation that more labour and resources are available beyond the immediate core of the colony. There is then a process of *integration* of the territory and its people and resources. Labour is persuaded, drafted and encouraged into the cash economy through a range of policies designed to make people need cash and begin to sell their labour. This requires migration to the main sources of employment, the cities and the plantations. The final transition sees the process of *disintegration* affect the colony and later the independent nation as families in remote rural areas send members in search of employment in order to pay hut taxes and school fees. This maps out on the territory of Tanzania as three distinct zones. The first is the zone of the export production economy, dominated by the production of goods for export. It includes the urban areas that act as trading, transport and administration hubs for these extractive activities. The second is just beyond the export production zone and is the zone of support, engaged in the provision of resources such as food, energy and water to the export production zone. Last comes the periphery which benefits very little from the core activities, but supplies seasonal or periodic labour. This maps a political economy on to the territory that suggests, rather than a process of modernisation diffusion, exploitation of Tanzania's people, resources and space by both Tanzanian elites based in the core and the colonisers and later the international markets.

Although these two contrasting approaches focus on Tanzania, they have been applied to other countries and to models of development throughout the world that saw contrasting roles for urban and rural areas. The Club du Sahel's recent research on urbanisation in West Africa points out that population growth in the region will increase the size of the cities and therefore the urban food markets. It is therefore necessary to adopt a positive approach to rural–urban interaction in order to ensure that this urban growth is sustainable. (For more on urban growth see Chapter 4 and for more on food systems see Chapter 2.) Earlier, the theories of W. W. Rostow (1960, 1990: 22–24) identified three major roles for agriculture in the transition between traditional society and the successful take-off of development. These are:

1 Agriculture must supply food. This is particularly important for urban-based industrial workers who have no access to the resources needed for food production, such as land and labour.
2 The extent to which agriculture can develop and earn foreign exchange may set the limit within which the transition to modernisation can take place. The development of agriculture provides both a supply for a modernising economy, for example in terms of food and industrial raw materials, and a demand for industrial outputs, for example for machinery and chemical inputs, which can further stimulate the process of industrialisation.
3 Agriculture must provide its surplus income to the modern sector. Rostow points out the importance of landowners to investment in the industrial revolution. Key to the land reforms of countries like Russia and Japan was the ability of the economy to raise investment, whether public or private, for the economic and social needs of the modernisation process.

Rostow's original account, published in 1960, suggests that economic growth progresses through a number of stages. This idea is centred largely on industrialisation and modernisation as the ultimate aims of development. Some researchers have criticised his work for being Eurocentric, because he appeared to assume that all development takes a similar pattern, each economy following a parallel path of modernisation towards a west European and North American model (Mehmet, 1995). However, Rostow illustrates the intimate links between the rural-agricultural and the urban-industrial in very clear terms. In the third edition of his book published in 1990, he suggests that the extent to which agriculture develops may limit industrialisation and ultimately modernisation. In addition, he suggests that the relationship between the two sectors may be interdependent. Therefore, in addition to the three roles for agriculture listed above, Rostow also argued that the income earned in developing agriculture may also be important to the transition to modernisation. '[F]or it is from rising rural incomes that increased taxes of one sort or another can be drawn – necessary to finance the government's functions in transition – without imposing starvation on the peasants or inflation on the urban population' (Rostow, 1990: 23).

The exploitation of the countryside by the city in the development process is very much the kind of extractive relationship between city and countryside developed by China. It is ironic that Rostow was explicitly anti-communist while China's analysis and development took place under a communist regime. This is also a key part of Myint's (1964) theory of *dual economies*. This idea identifies two techniques of production, one traditional and one modern. Myint's dual economies comprise basically Western capital-intensive production and 'backward' rural-based labour-intensive sectors. This is not far from Slater's (1974) account of the three zones of the colony, though he interprets the spatial patterns quite differently. Followers of the dual economy analysis were often explicitly pro-capital, and by implication pro-metropolis, and anti-rural. According to Mehmet (1995), they often characterised the rural areas as having a high labour surplus, and the modern urban sector as the engine of the economy. A consequence of this kind of analysis was the belief that the rural areas were the source of underdevelopment in the developing world, implying that investment for development should be targeted at the productive sectors where the greatest return was possible, the urban and industrial sectors. For Mehmet (1995) one of the main objections to this idea was the underlying assumption that the rural producers were 'irrational' because they were unproductive. These aspects of the links between the rural and agricultural sectors of the economy suggest that it is important to study them so as to understand the process of development itself.

Development theory, and research on the process of urbanisation, highlight polarisation of societies in relation to urban and rural, traditional and modern, industrial and agricultural. Some theorists argue that development is about progression from one state to another. In the past this has been tied in with the urbanisation of rural areas. Rostow was one of the most influential of these modernisation theorists, suggesting that, based on research on the modernisation of European markets, all economies aspired to the condition of 'high mass consumption'. A simplified version of Rostow's model of transition to modernisation is illustrated in Table 1.6.

Rostow argued that all economies began from a stage of low productivity, which he described as the 'traditional society'. He suggested that societies progressed through a number of stages at various speeds to a final stage of 'high mass consumption' (see Table 1.6). He conceptualised traditional society as one with rudimentary production functions, 'based on pre-Newtonian science and technology and on pre-Newtonian attitudes towards the physical world' (Rostow, 1990: 4). Such societies are highly agricultural,

Table 1.6 ***Rostow's stages of economic growth***

Stage	*Economic features*	*Implications for rural–urban interaction*
Traditional society	Agriculture dependent	Most production subsistence or feudal
Preconditions for take-off	Loss of power of landed elites	Growing settlements offer opportunities for commercial agriculture and some migration; generation of investment through growth in agriculture and extractive industries
Take-off	Increasing focus on industry and commercial agriculture	Growing cities lead to increased demand for food and raw materials; improving transport and communications; rapid rural–urban migration
Drive to maturity	Diffusion of production and transport technology	Transformation of agriculture and industry through introduction of technology; development of transport links into networks; increasing importance of links between agriculture and industry
High mass consumption	Majority earn beyond basic needs; focus on durable goods	Focus on consumption, reduced influence of agriculture
Beyond mass consumption	Increasing proportion of workforce self-employed and growing numbers in information and media industries; high levels of disposable income	Growth in global networks of communication and interaction

Source: adapted from Rostow (1990).

largely dependent on agriculture and with social structures that are essentially agrarian.

One of the crucial stages, according to Rostow, is the third stage of growth which he described as the 'take-off' stage, using the analogy of an aeroplane beginning at a slow speed and having to achieve a take-off speed before its final ascent to cruising altitude. Before achieving take-off, Rostow argued, it was necessary to have in place the 'preconditions for take-off'. These preconditions include changes in attitude and political and economic structure which prepare the economy for investment in infrastructure and a widening of economic activity both within the nation and beyond, in industry and in agriculture. This implies both the demise of the power of the traditional landed elite, which Rostow sees as blocking development, and the emergence of a modern nation state which is intent upon progressing to modernisation. It also implies a shift in power base from rural-based elite interests to an urban-based political elite, increased rural–urban trade and increased migration. These changes first occurred in western Europe, as developments in modern science began to be translated into improved restructuring and efficiencies in production in both agriculture and industry, periods of change referred to as the agricultural and industrial revolutions. Rostow seems to suggest that Britain first achieved the preconditions for take-off by accident of history and geography. The more common pattern is for a society's transition to take-off to begin as a result of external initiatives, such as colonisation, the implication being that modernisation was a process of diffusion triggered first by colonisation and later by international trade, aid and investment links.

Rostow describes the crucial 'take-off' phase as 'the great watershed in the life of modern societies' (1990: 7). During this phase the old blocks to progress such as low levels of savings and investment, high levels of agricultural dependence and slow industrialisation are overcome and the forces for economic progress come to the fore. New industries expand rapidly, generating other industries which supply and service them and their labour force directly. New methods of production become more widespread in both agriculture and industry. Agriculture becomes more market oriented, making its contribution to the development of industry and the economy through the three links described earlier in the chapter, first supplying food, second generating food, raw materials and demand for manufactured goods for inputs and, finally, supplying surplus to the modern sector in the form of goods, labour and investment.

Once take-off has begun, the next in Rostow's stages of economic growth is a long, sustained 'drive to maturity', as the developments in industry and technology are diffused throughout the society, integrating all parts of the economy to contribute to and benefit from the economic growth that has occurred. In relation to urbanisation, Rostow argued that the 'take-off' stage 'sets up a requirement for urban areas, whose capital costs may be high, but whose population and market organisation help to make industrialisation an on-going process' (1990: 58). Rostow implied a change in the nature and role of cities since clearly urban areas pre-date the 'take-off' stage. So the development of settlements is seen by Rostow as a necessity for economic growth, implying that modernisation will diffuse outwards from the core city into the rural hinterland and beyond.

Once this phase is complete, incomes rise to a level where the majority earn enough to go beyond basic needs into a phase of 'high mass consumption'. Economies begin to focus on consumer durables, such as bicycles, sewing machines, various household electrical goods and, perhaps most important, cars. Societies begin to consider issues of social welfare and security. In the third edition of his thesis, Rostow speculates on what comes beyond this stage. This resulted in the addition of a fifth stage, illustrated in Table 1.6, in the row titled 'Beyond mass consumption'. In this stage Rostow suggests that there will be an increase in the flexible working arrangements facilitated by improvements in communications technology.

In Rostow's model transition from a traditional to a modernised society there will inevitably be a change in the relationship between the rural-agricultural sector that dominates the traditional society and the urban-industrial sector that dominates the high mass-consumption society. Some of Rostow's critics suggest that he is not only pro-industrialisation but also biased against rural-agricultural sectors (Harvey, 1989). It is hard not to arrive at this conclusion in light of his suggestions that the urban–rural links will increasingly favour the urban-industrial sector and that this is an important precondition for the 'take-off' stage. According to the model, agriculture is exploited to provide for the needs of an industrialising and urbanising society. Under such a transition, then, not only are the rural–urban interactions in the developing world increasingly exploitative of the rural-agricultural sector but this is a necessary precondition for economic growth. According to Rostow's analysis, this is a positive development for society in general.

A more recent approach to this question, focusing in particular on the agricultural-to-industrial employment transition, is that put forward by Tiffen (2003) and illustrated in Figure 1.3. The model is based on 'some of the most basic and durable concepts in economics' (Tiffen, 2003: 1345), including:

- the division of labour by specialisation, leading to improved productivity and technological improvements;
- the accumulation of population, facilitating the exchange of services, goods and information;
- the division of labour being constrained by market costs in relation to market values;
- productivity improvements requiring investment of either labour or capital and diminishing returns on this investment until a change in technology or in product take place.

The result is a process of change in the employment structures of an economy over time. Tiffen suggests that the periods of change can be divided into three. In phase A almost all labour is applied to subsistence food production. The population is sparse and scattered and so surplus is rarely sold on. In phase B, or the transition phase, access to external capital facilitates the expansion of the non-agricultural sector and the transport infrastructure. 'Change accelerates as productive technologies and technologies for the exchange of goods and information improve. Concentrations of people facilitate the exchange and development of ideas' (Tiffen, 2003: 1347). This change and the increasing contact with cities and international markets give farmers opportunities to benefit by producing surplus. This surplus in turn promotes the development of the consumer goods and services sectors of the economy that are largely located in cities. The introduction of improved technology and methods of farming releases labour to work in the urban industrial and services sectors, which in turn stimulates greater markets for agricultural goods. Some regions have never benefited from the appropriate conditions. For example, in many Latin American countries the structure of land holding meant that the benefits of export agriculture accrued to a small elite, creating insufficient stimulus to the services and consumer goods sectors. In Africa, major transport infrastructure investment was patchy until the late twentieth century. In phase C the internal market becomes very large, supporting a manufacturing and services sector that is no longer dependent on links with agriculture. As this sector grows and incomes increase, the range and diversity demanded in agricultural produce also increase. The phases have no fixed time period and Tiffen suggests that phase B could take between 50 and 500 years, with the possibility of moving both up and down the gradient depicted in the model in Figure 1.3.

A similar model of agricultural transition is presented by Hill (1997; see also Robinson, 2004, and Rigg, 1998a, 1998b) for South-East Asia. In this, Hill suggests that there is a productivity gap which occurs during the transition as the number of farmers increases, but the proportion begins to decline, resulting in periods of shortage in agricultural labour (even in densely populated regions) before the introduction of improved techniques and mechanisation. An alternative analysis of the relations between rural and urban realms focuses on the way that 'rural' and 'urban' are interpreted themselves. For example, some radical theorists (such as Slater,

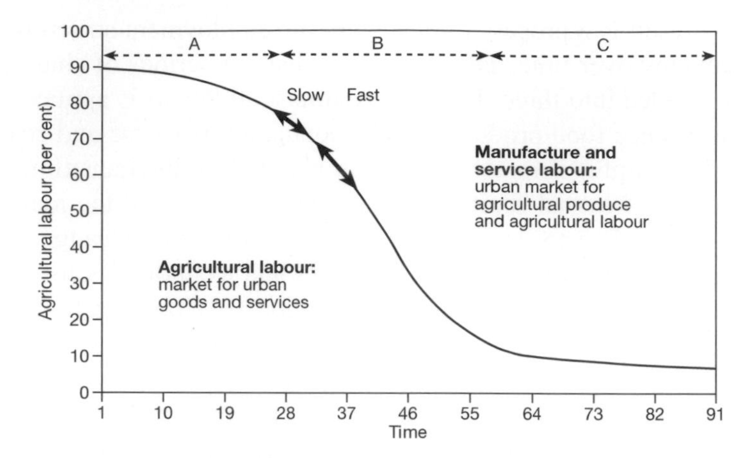

Figure 1.3 ***Tiffen's model of changes to agricultural and service and manufacturing labour over time.***

Source: reprinted from *World Development* 31(8), M. Tiffen, 'Transition in sub-Saharan Africa: agriculture, urbanization and income growth', pp. 1343–1366, © 2003, with permission from Elsevier.

1974) have argued that cities are the gateways through which the developing countries have been and are being exploited. The counter-argument to this, of course, is that cities are the gateways through which the rural areas access international markets and international investment and aid.

Since the end of the Second World War a number of national-level initiatives have been rolled out by urban-based governments in an attempt to achieve rural development. These have incorporated rural dwellers' needs and priorities to varying degrees. For example, early initiatives focused on the restructuring of the organisation of the main means of production, labour, land and capital, to take advantage of the 'trickle-down effect' from industrialising urban core areas. Examples of these approaches include rural production sector reform, land reform and village resettlement. As these policies were found wanting and the expected benefits of urban-based development did not find their way to the rural peripheries, there was a shift towards integrated rural development approaches which emphasised the requirement to meet the needs of the rural dwellers, such as nutrition, water, health care and transport. However, as large amounts of money were invested in this approach with limited results, the focus shifted to reducing state involvement in development initiatives and attempting to encourage the private sector. During the 1980s there was a shift in interventions and government policy back towards more market-based approaches. These emphasised the need for smaller, locally oriented approaches.

Evidence suggests that the sustainability of urban and rural areas appears to be intimately linked (Rees, 1992; Main, 1995; Hardoy *et al.*, 2001). Whether considering the environmental or economic sustainability of the rural population, or the food security and energy needs of the urban population, most research finds links across the urban–rural divide. Despite this, there is little evidence of policy makers appreciating the importance of these links. Only recently are international and other aid agencies developing interventions which show an acceptance of the importance of these links. Discussion later in this book includes examples of rural and urban dwellers developing survival strategies that bridge the urban and the rural realms. Indeed, the different benefits and costs of urban and rural areas provide precisely the opportunities that multi-locational and migrant households are seeking (see Chapter 4). The next section briefly discusses the theory on the role of cities and the relation between cities and the rural areas.

As a result of this discussion it can be seen that the relationship between city and countryside in developing countries is qualitatively different from that between city and countryside in the developed world. In the world's poorer countries there is a higher dependence on rural activities for wealth and employment, for example in agriculture, mining and fisheries. By contrast, in the developed world most wealth is created in the urban areas. This has implications for the power relations between the city and countryside and results in overlapping and conflicting concerns.

The shift in emphasis between urban and rural

While rural and urban dwellers exploit urban–rural differentials as a strategy for enhancing their livelihood options, a longer-term process is under way in developing countries which creates additional opportunities and threats to households which are urban, rural or straddling the continuum. This process is urbanisation. Urbanisation is often conceptualised in terms of shifting demographic structures, through migration, rapid natural urban growth, or physical city expansion. However, urbanisation is also conceptualised as the spreading influence of the city on rural areas. This takes the form not only of making bits of the country into cities through building, but also of extending the activities and influences of the city into the countryside and making what happens in the countryside increasingly influenced by the needs of the city. The main consequence of extending urban influence into rural areas is that relations are altered over time and under different influences. For example Potts (1995) and Jamal and Weeks (1988) illustrate the different influences of structural adjustment policies on migration and urban bias in rural and urban areas. These are discussed further in Chapter 4.

Previous work on rural–urban interactions

McGee (1991) describes an Asian version of an urban conurbation which he calls a *desakota*, derived from Indonesian (*desa* – village; *kota* – city). This combination of city and village describes a complex area of urban and densely settled, urban-influenced rural hinterland. This hinterland shows signs of close interrelationships with the urban economy. McGee also describes a process of *desakotasi*, or the dynamics of the development of the *desakota*. The interesting element in this theory is the idea that the distinction between urban and rural is without purpose since the interconnections between their populations are more important than the differences between them. The process of *desakotasi* has only been identified to a limited extent outside Asia (see for example Simon *et al.*, 2001). This idea of the continuum between what is urban and what is rural is suggested in an earlier model provided by McGee (1991) and illustrated in Figure 1.4. In this he suggests that it is possible to conceptualise the urban and rural as being on a continuum. The development of urban areas is seen as inevitable. What varies is the speed of the development process. This creates urban and rural differences. However, the model suggests that these will ultimately even out. The means by which this will occur are the links between the two realms. For example, as economic development takes place and urban areas develop ahead of rural areas they will attract the unemployed through rural–urban migration. This will eventually free the rural sector from the burden of surplus labour. This does not, however, take account of the economic crises of the 1970s and 1980s when, some researchers suggest, certain urban and rural areas actually reversed their development relationship (Potts, 1997). However, this is discussed in more detail later in the book.

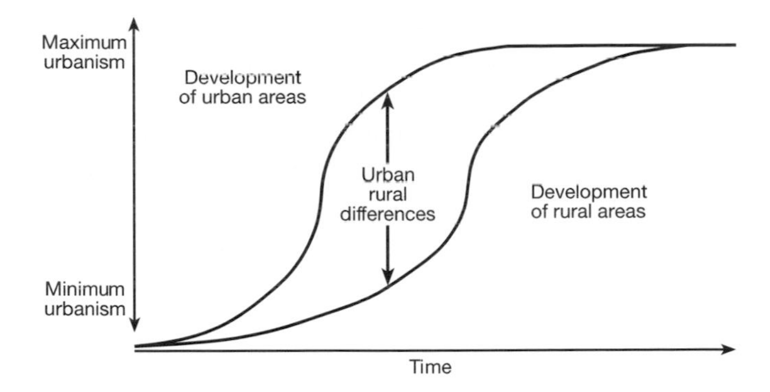

Figure 1.4 ***Simplified process of modernisation.***

Conclusion

This chapter has set out the broader context in which the research on rural–urban interaction has taken place. This has involved discussion of a number of key themes relating to the causes and consequences of this interaction. One key debate in the literature focuses on whether the relationship between the city and the countryside is a benign or exploitative one. This debate links this area of research into the heart of development theory where there has long been a focus on whether core–periphery relations are positive or negative. The debate usually focuses on the consequences of the very rapid urbanisation and urban growth experienced by the countries of the developing world. A second key issue is the problem of defining the territories under study as 'town', 'city' and 'countryside' and the divide between – especially in countries where cities are growing outwards very quickly. A third issue is that cities in the developing world often form the gateway between the rest of the country and the international markets. The evidence suggests that they rely more heavily on their own hinterlands than do cities in the developed world, so their relationship with their hinterland is very important. This link is reciprocal, as the cities of the developing countries provide the rural areas with their links to world markets.

Discussion questions

- What are the main things that research has told us about rural–urban interaction?

- To what extent have urban areas contributed to the development of rural areas and have rural areas contributed to the development of urban areas?
- Review the international agency approaches to rural–urban interaction. Try to find examples from the websites of two agencies which are acting on their intentions.

Suggested reading

Binns, T. (2002) Dualistic and unilinear concepts of development. In V. Desai and R. Potter (eds) *The Companion to Development Studies.* Arnold, London. 75–80.

ODI (2002) *Rural–Urban Linkages*. Key Sheets for Sustainable Livelihoods No. 10. Overseas Development Institute, London. Available at http://www.odi.org.uk/keysheets/ [last accessed: 28 November 2003].

Potter, R., Binns, T., Elliott, J. and Smith, D. (2004) *Geographies of Development.* Pearson, Harlow. Second edition.

Rigg, J. (1998b) Rural–urban interactions, agriculture and wealth: a southeast Asian perspective. *Progress in Human Geography*. 22 (4), 497–522.

Tacoli, C. (1998a) *Bridging the Divide: Rural–Urban Interaction and Livelihood Strategies.* Gatekeeper Series No. 11. International Institute for Environment and Development, London.

2 Food

Summary

- **One of the most important interactions between the urban and the rural areas is the provisioning of food.**
- **Because of food's central role in human life, a considerable body of research has built up around the issue of food systems.**
- **One aspect of the blurring of the urban–rural divide is the increasing cultivation of food in and around the city, known as urban and peri-urban agriculture.**

Introduction

This chapter focuses on the theme of food in rural–urban linkages. It argues that research on food supply to cities provides tangible evidence of the more difficult concepts of power, economy and society that structure rural–urban relations in the developing world. By the end of the chapter you should have a reasonable understanding of the importance of food to both cities and rural areas. The chapter introduces some of the ideas presented in research on rural production for urban areas, urban food supply systems, including food chains and trading systems, and urban and peri-urban agriculture. It concludes by considering how it may be possible to integrate these different aspects of rural–urban food supply and how this may help plan for sustainable urban *and* rural development in the future.

Understanding urban food supply

The study of the supply of food from rural areas to the city involves an understanding of broader social and economic as well as geographical processes at work in rural–urban interactions. The city, as an area of considerable food deficit, serves to focus the trade of food within the country. In this way the city serves as a focus for food distribution in the same way as it does for the political and economic life of the country. For example, Guyer (1987) discussed food marketing for cities in Africa in terms of food chains, thinking about food supply by focusing on the links from producer to consumer. She argued that these food chains are important not only for their functional role:

> [t]hey are also organisations rooted in articulated social and economic structure. Throughout the continent, they form a bridge between conditions of production in an African society and ecology and the conditions of exchange and power in a national and international political economy.
>
> (Guyer, 1987: 6)

In the case of urban food supply they are also a key conduit through which the relations between urban and rural areas are articulated. The study of urban food supply in less developed countries (LDCs) is therefore of crucial political and economic significance.

In global terms one of the most emotive concerns about the future of the poorest countries in the world is ensuring food security. Much of this concern centres on countries with rapidly growing populations and their strategies to increase food production at a rate faster than population growth in order to ensure the future sustainability of food supplies. This involves two apparently contradictory aspects. The first is that many of the developing countries are heavily dependent on agriculture and, in particular, food production. The second focuses on the problem of food security, food shortages and starvation. Koc *et al.* (1999) estimate that around 35,000 people die each day from hunger. Meanwhile in cities, fiscal and foreign exchange constraints resulting from structural adjustment, together with rapid urban growth, have created particular challenges for maintaining urban food security (von Braun *et al.*, 1993). In particular, the evidence suggests that where economic deterioration was at its greatest during the 1980s, the urban poor were particularly hard hit and this had an impact on their ability to provide household food security.

Potts (1997) reports research that in the late 1980s the average monthly income for an urban Tanzanian household of six would cover their food costs for only three days. Binns and Lynch (1998) reported that in Nigeria in 1997, when the country was experiencing a particularly challenging economic situation, the chairman of the Federal Economic Recovery Commission accepted that the average civil servant's take-home pay would hardly take him (*sic*) home. Similarly in Asia during the economic crisis of the early 1990s, the urban population suffered from a cost–income squeeze, as the cost of living increased and the value of earnings and employment declined. Beall (2002) argues that urban dwellers have a range of strategies for dealing with economic crises and austerity. These include risk-spreading behaviour such as *allegamiento*, a practice common in some parts of Latin America of co-residence of different household units. In sub-Saharan Africa a common response to economic restrictions is to reduce the number of dependants by having a child fostered by the household's village-based relatives. This indicates the importance of keeping a link to the rural-based family. An alternative, where child labour is in demand, is to send children to work to bring in additional household income.

Structural adjustment policies resulted in the removal of food subsidies and the retrenchment of government and parastatal employees. These impacts resulted in considerable protest from the residents of cities in the countries concerned. In some cases, such was the level of dissatisfaction that the removal of food subsidies in particular resulted in civil unrest in the form of what was described as 'food rioting' in the capitals and main cities. Such riots took place in Tunisia (1984), Zambia (1986), Nigeria (1989), Morocco (1990) and Jordan (1996), and there is some evidence that increased food prices were one of the triggers for the Tiananmen Square demonstrations in Beijing in 1989 (Brown, 1997). As Sporrek (1985: 12) puts it: 'Few conditions can so easily deteriorate into a situation of social unrest as a malfunctioning urban food system.' Urban food supplies have crucial political implications, as mentioned above. In the case of both Tunisia and Zambia the food riots have been linked to the raising of staple food prices in agreement with IMF recommendations (Bryceson, 1985; *Observer*, 14 December 1986). In each of these six examples of food-related riots referred to above, the urban unrest placed enormous political pressure on the government concerned. The importance of this is explained well by Pletcher in his examination of Zambian agricultural policy.

> Social power for the state lies with those who can topple it. In the short term farmers offer little of the immediate threat which strikers or angry crowds present. Therefore, when tight times come, it is easier politically to squeeze the farmer a little harder than the rest. The long run effect is to make farming so unattractive it deteriorates.
>
> (Pletcher, 1982: 220)

This is a stark illustration of the differential power balance between urban and rural dwellers, where fledgeling democracies are reliant on relatively straightforward links between food supplies and votes.

The deteriorating context is compounded by the lack of research in the area of urban nutrition. As an indication of this von Braun *et al.* (1993) report that of 1,324 English-language publications issued by the Institute of Nutrition of Central America and Panama between 1949 and 1985, only 35 were related to urban communities, and only three exclusively to urban nutrition. The authors go on to speculate that the urban-bias thesis was influential in research on rural poverty and nutrition, favouring conclusions that see rural problems as being related to urban bias in national policies and in private sector activities. Paddison *et al.* (1990) suggest that the development and viability of cities in developing countries depends on the availability of sufficient food supplies for their populations. However, the deficiencies of the literature on food supply are highlighted by Drakakis-Smith (1991: 51): 'The sparse nature of research of this topic stands in marked contrast to the literature on housing which is more voluminous, and yet urban social protest is far more likely to be initiated by food shortages or price rises than by housing problems.' There is research on public distribution systems in a range of countries. For example, Ramaswami and Balakrishnan (2002) found that food subsidies in India were ineffectively targeted at the poor, but to remove them would have had adverse effects on low-income groups. They argued that a reduced and controlled procurement approach combined with reduced subsidy impacts by allowing a greater supply to reach the market, thus reducing prices. Other examples in the literature include Mexico (Roberts, 1995) and China (Croll, 1983), but much of this fails to address the issue of the rural–urban divide and the implications of public procurement for producers.

This, however, is in contrast to the importance for rural areas of the production and trade of food for cities in small, fast-growing economies. Most developing countries rely heavily on agricultural

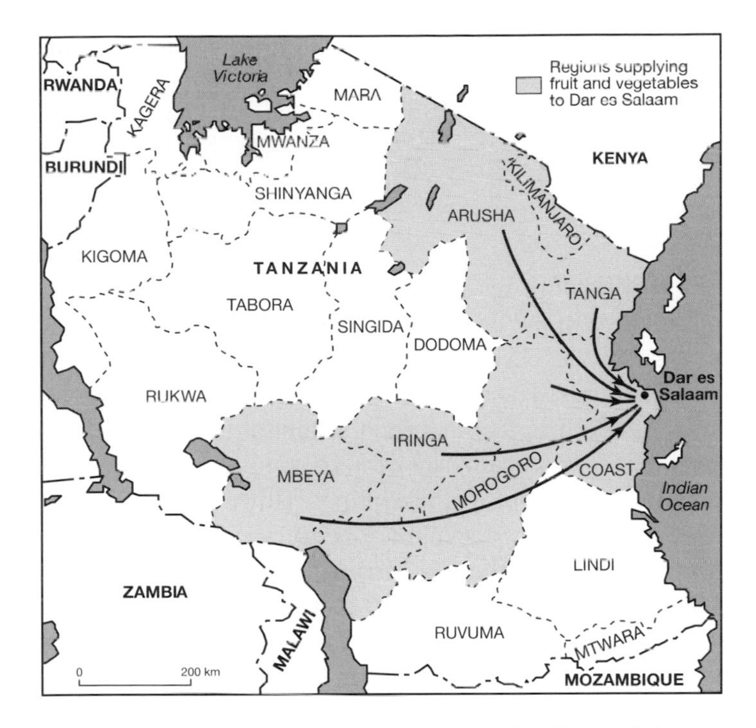

Figure 2.1 ***Main regions supplying fresh food to Dar es Salaam.***

production for their economic growth (see for example the discussion of the links between agriculture and industrialisation in Chapter 1). Much of this agriculture is focused on food production. The map in Figure 2.1 illustrates the growing geographical significance of the supply of food to the city of Dar es Salaam, Tanzania, where it is driven in from very great distances. The map illustrates the expanded area that supplies the rapidly growing city. This is evidence perhaps of an element of what has become known as the city's 'ecological footprint' (see Chapter 3 below). However, this map masks a division in supply chains between the commercial and the small-scale farmers, and between the wealthy and the low-income urban consumers. This contrast was found by Mosley (1983) in colonial Rhodesia, now Zimbabwe. He reported that during 1921, the European commercial farmers drove down the price of maize to 10 shillings, puncturing a boom which had brought a measure of prosperity to the mainly small-scale African producers. Well after independence in Zimbabwe, Drakakis-Smith (1990) found evidence that a divergence in access still existed, both between different groups of urban consumers to the food supplied to the city and between white and black producers to different food chains supplying the city. The situation in Zimbabwe has recently deteriorated, in part as a result of political agitation around such contrasts.

Food supply systems

A small amount of research has focused on networks which supply food to cities in the developing world. During the late 1980s a literature on this began to emerge which uses a range of approaches to food systems (Hansen and McMillan, 1986; Gittinger *et al.*, 1987; Guyer, 1987). The fact that food is central to human life and society means that food systems are of interest to a very wide range of academic approaches and disciplines. However, previous approaches were constrained by disciplinary boundaries, which left studies open to the criticism that they had arrived at erroneous or simplistic conclusions because of an incomplete analysis of the food systems under consideration. For example, an economic approach to rural markets may demonstrate that it is not profitable for many of the sellers to attend them. This conclusion does not take into account the immense social and cultural importance of such markets, which can make a purely economic gain less central to the motivation of those who participate. Jones (1972) describes this phenomenon as one of African markets' malfunctions, whereby 'participants in the market are unproductively employed or are not seeking economic gain. Many social, political and administrative activities, in addition to the purely economical ones of buying and selling, typically characterise each marketing meeting' (Jones, 1972: 10). Indeed, one female trader interviewed by the author in Lushoto District, Tanzania, described how women may purchase produce from a neighbour in order to be able to participate as sellers in the periodic market, even if this meant they would be out of pocket at the end of the day. The reason for this was the social and cultural importance placed on attending the markets.

It may also be argued that quantitative research in such situations is very difficult. It is extremely difficult to monitor accurately the volumes, prices and transactions involved in such rural periodic markets. The units of measurement are not uniform, with sellers frequently selling produce in arbitrary piles or in containers which vary in size depending on the commodity and its condition and on the relative negotiating skills of buyer and seller. In the protracted discussions over price that can often take place in such markets across the developing world, whether rural or urban, both parties may suggest different rates of exchange, varying both the price and the volume. In addition, preferential rates may be offered to, or obtained by, customers known to the seller or those purchasing larger quantities, and increased rates to those who appear able to afford a higher price. Additional variations in price may be obtained by either the seller or the buyer according to the variety, origin or quality of the produce. Finally, many food crops in Africa are produced on mixed-crop plots, where trees may be interspersed with one or two other crops. This provides a variety of vegetative input to the soils while extracting different nutrients, and the trees provide shelter for the lower crops from both direct sun and tropical rain storms. This makes the estimation of crop yields per unit of land extremely problematic. Ways round this may involve adopting a range of research methods that include, but do not rely on, quantitative data, suggesting an interdisciplinary approach is required.

Food systems themselves – due to the importance of food to survival – are the subject of study in such diverse disciplines as agriculture, anthropology, economics, geography, government policy, marketing, medicine, sociology and, doubtless, more besides. The result is research that approaches food supply systems from a range of perspectives and in a way that comes closer to understanding the realities of the 'confusing complexity of people, families, communities, livelihoods and farming systems' (Chambers, 1997: 54). This requires a shift of focus towards the people involved in, and affected by, the food systems, rather than towards aspects that favour disciplinary insights. This is not to detract from the insights provided by different approaches, but to foreground ways of applying these insights for the benefit of the stakeholders in the food systems. 'In studying market place trade as an aspect of economic development different social science disciplines have focused on different aspects of the problem' (Epstein, 1982: 209). For example, anthropology approaches its subject from the perspective of the groups of people involved in a system, requiring the researcher to visit and live in the rural producing area, making observations and interviewing the participants. Economics, on the other hand, has been more interested in variations in price and supply and the role of capital in a system and will collect data measuring these factors, or at the very least approach the subject from this point of view. These approaches are legitimate within their respective disciplines, but they do not provide an overall view of the food supply system. Guyer (1987: 6) argues that this diversity has provided researchers with varied insights into food supply; however, a juncture has been reached in its study where 'particular dynamics cannot be studied without a re-composition of the field'. Drakakis-Smith (1991: 51) comments that research on food supply

> in geography, but also in related social sciences, has been patchy so that although some elements, such as the nutritional problems resultant from inadequate diets or the operation of markets, are reasonably well covered, it is difficult to formulate a comprehensive framework.

Because of its highly complex function in human life, 'a broadly based integrated policy approach to supply, distribution and consumption is essential if everyone is to be assured access to food' (Gittinger *et al.*, 1987: ix). Previous attempts have been limited by their narrow approach. In his book *Marketing Staple Food Crops in Tropical Africa*, Jones (1972) took up the subject from an economic perspective, concentrating on an analysis of four African marketing systems. He compares each of these systems, measuring their performance using variables of price and volume, against his idea of a free market, focusing on what he calls 'imperfections' in the African market systems. As a result, the conclusions which Jones reaches are grounded in this way of examining food marketing systems. This may result in problematic conclusions unless backed up with additional data. Price data alone are open to interpretation, as discussed above, and are, therefore, insufficient for the purposes of policy analysis without additional evidence. For example, the result of an analysis may show the price stability of a commodity. However, without additional evidence this could equally well be the result of either monopolistic control or competitive conditions. There is no way of deducing a conclusion where there is more than one equally likely explanation, without interpretation of sociological or political data. In addition, serious questions may arise regarding such secondary data as are available for wholesale and retail activities. Frequently they concentrate on the formal structured market and ignore the unstructured informal market. Samiee (1990: 36) reports that, according to 1980 UN data, Egypt, with a population of over 41 million, has only 1,018 retail establishments.

Lynch (1992) puts forward a typology of five approaches to the study of urban food marketing systems in Africa (see Table 2.1). Each approach focuses on a particular theme. These categories are based

Table 2.1 ***Categorisation of urban food systems research***

Theme	*Approach*	*Focus*	*Limitations*
American	Economic decision-making	Price and cost analysis of commodities	Fails to integrate historical dimension into analysis
Francophone	Ethno-geographic	Social organisation of delivery of 'consumption basket'	Lack of focus and simplification of 'public' and 'private' systems
British	Economic sociological	Interaction of food supply with standards of living	Historical dimension weak
Historical	Regional social history	Local historical changes in institutions and entitlements of food supply	Raises questions, stops short of policy recommendations
Spatial	Spatial market	Patterns of spatial distribution and networks of markets	Limited consideration of economy and society unless contributing to spatial pattern

Source: adapted from Guyer (1987) and Lynch (1992).

to a large extent on Guyer (1987), who posits four approaches: a decision-making framework and economic sociological, ethno-geographic and regional historical approaches. Aragrande (1997) suggests seven categories as an alternative. In addition to Guyer's, he suggests that there is a legal and an evolutionary approach. He also suggests a development of the evolutionary approach that appears to provide a historical dimension to classical economics, giving it a concern with innovation. The legal approach is to a large extent subsumed by Guyer's regional social history approach which gives considerable attention to institutional change through time. Guyer's (1987) loose categorisation therefore forms the structure of the following brief discussion of research on the relationship between cities and the countryside through food networks. To Guyer's four approaches Lynch (1992) adds a fifth to take account of the spatial dimension of markets and market networks. These approaches are now considered in turn.

Economic decision making approach

Jones (1972) examines the workings of the market and the ways in which government can intervene to improve its efficiency, such as disseminating price information (see Chapter 5 below for a discussion of market information systems), forming commodity exchanges (he quotes the success of such exchanges in the United States of America) and geographical production specialisation. Guyer (1987) reviews a series of works associated with Jones and the Stanford Food Research Institute, which put forward price analysis as a method of research, and which employ historical and sociological evidence to strengthen price analysis interpretations. However, Guyer argues that the integration of historical and sociological study is indispensable to a comprehensive food market analysis, and that rather than being peripheral, this evidence should be examined in conjunction with quantitative price data. There are two reasons for elevating qualitative data to a status similar to that of economic data. First, although qualitative data may be considered less objective, questions may also be raised regarding the accuracy and 'representativeness' of the price data. Indeed, the immense difficulties of collecting such data in developing countries increase the necessity of handling them with extreme caution. Second, Sen (1982: 159) argues that there is a 'need to view the food problem as a relation between people and food in terms of a network of entitlement relations'. In other words, the political dimension is fundamental to an understanding of food supply problems and food security. Such factors are not readily quantifiable, particularly in the context of a developing country.

Polly Hill (1963) demonstrates the difficulties of approaching the subject of markets in Africa from such a quantitative point of view (though her comments could equally apply to India, Pakistan, Brazil or any other country where markets are a key element in urban food supply chains):

> It is not so much the heat, the glare, the bustle, the over-crowding, the noise, the shouting (and consequent hoarseness), or even the sneezing caused by open bags of pepper and maize (for all this is compensated by the very courteous behaviour of Africans in markets) – the difficulties are rather the extreme fluidity and complexity of the undocumented situation and the need to trouble informants at their moment of maximum anxiety, when they are concluding transactions. Perhaps one day economists will devise new techniques, presumably involving teamwork, for tracking and identifying the variables in these anti-laboratory conditions. Meanwhile it is to be hoped that the hardships endured in market places by individual anthropologists will not have been in vain.
>
> (Hill, 1963: 445)

One of the reasons for this complexity in market places is that they are an arena in which urban and rural come into direct contact. They are the central rural–urban interface in food chains, whether they are urban-based, drawing on a range of rural areas to supply a city's retail and wholesale markets, or rural markets that draw traders from a number of cities and farmers from the surrounding area.

Ethno-geographic approach

The economic decision-making approach is associated by Guyer (1987) with themes of research that were dominated by academics from US institutions. She suggests that there are two further approaches that are strongly influenced by the countries from which most researchers originated, the 'Francophone themes' and the 'British themes'. She points out that their proponents are by no means confined to these nationalities, but the approaches seem to have arisen out of the different perspectives of the researchers and the different government priorities, both partly as a result of the distinct historical experiences of Britain and France. For example, the French colonised large, sparsely populated countries in the Sahel and Central Africa, whereas the British governed densely populated areas such as the Gold Coast and Nigeria. In addition, the distinction in these approaches may have come out of the different national histories of the colonial powers themselves. There is a long history in France of criticising the government use of control of staple food supplies to Paris, creating tensions between the state, the consumer, the producer and the intermediaries. In Britain, on the other hand, wage unrest after the Second World War resulted in the introduction by government of wage subsidies. Furthermore, French scholars have tended to approach the question of food supply as a highly colourful anthropological phenomenon, describing *how* food is traded, whereas British scholars have tended to approach the question from the point of view of to *whom* food is supplied , as exemplified by Sen in his book *Poverty and Famine*:

> if we look at the food going to particular groups, then of course we can say a great deal about starvation. . . . one is not far from just describing the starvation itself . . . clearly, they didn't have enough food, but the question is *why* didn't they have enough food? What allows one group rather than another to get hold of the food that is there? These questions lead to the entitlement approach.
>
> (Sen, 1982: 154)

The French tradition, which approximates to an ethnographic approach, concentrates on the different ways of interpreting the social organisation of local marketing systems which deliver goods. As such, this school focuses on certain issues, such as the sources of the food, the types of transport, the locations and number of transactions. This approach also makes certain assumptions, such as 'the institutions of distribution take precedence over market principles as the subject of research' (Guyer, 1987: 15). The state, in this case, becomes not the initiator of beneficial policies for the workings of the market, but simply another social institution, which is potentially, and often actually, in conflict with indigenous organisations or population groups. However, there are problems with this approach. It is swamped with enormous and graphic descriptive detail, but can lack focus. This approach is exemplified in the work of Piault (1971), Cohen (1971) and Amselle (1971) describing aspects of modern-day West African markets. There are questions left unanswered concerning the definition of research priorities and the most important issues involved. The means by which the institutions involved in food supply have influenced policy and prices are rarely covered. Finally, there appears to be no generalised framework or development of strategies for further investigation.

Economic sociological approach

The approach, associated by Guyer with British researcher perspectives, may be described as a consumer-oriented one, focusing on urban real incomes and poverty in relation to food supply. It links prices and marketing institutions to real incomes and standards of living, investigates the access of lower-income households to food markets and the sources of their supply, and examines social class and local social structures and their impact on food distribution and consumption. Its focus is on social welfare and, although it may analyse production, distribution and consumption, it will always return to the income implications for the producer, the trader and the consumer. This is most clearly explained by Croll (1983: 7), in a study of food supply in China, when she says it 'examines both access to national and local food circuits and the demands that food policies make on the material and labour resources of the household so that it produces, purchases and processes sufficient food for its members' subsistence'. This concentration on class (or occupation as an indication of class) allows questions of the social implications of the economic process to be addressed, and brings the issue of poverty into focus. There is a concern with the role that low incomes play in entitlement to food and the means to obtain food. For example, it may view state intervention in terms of economic welfare, and therefore political stability, rather than market efficiency (Sen, 1982; Croll, 1983; Ellis, 1983). This approach appears to satisfy Guyer's (1987) demand for a historical dimension by referring to post-independence or immediate pre-independence experience, although it tends to be limited in this by the short time period over which such data have been collected. This approach neglects the question of rural–urban interaction or bias because of its concentration on class. Its objectives are narrow, though generally explicit. 'It is concerned with how acceptable, or at least liveable, income levels have been established in different areas, under different conditions' (Guyer, 1987: 18).

Regional social history approach

The approach adopted by Guyer in her own work may be termed a historical approach. It attempts to incorporate some elements of the above three traditions into a historical approach to urban food supply analysis. Guyer argues for the integration of 'the three dimensions which have tended to be separated in Africanists' work: market prices, the social and political organisations of production and trade, and the class implications of incomes' (Guyer, 1987: 19). She argues later that a focus on local areas, that is on cities and their food-supplying hinterlands, where all these dimensions meet, is one step towards building an understanding of the organisation of a society, by allowing it to be examined on a small scale. Guyer suggests that '[u]nderstanding the influence which former organisational rubrics and material conditions exert throughout a particular region's . . . [history] . . . is an agenda with which . . . case studies engage, but which is, ultimately, a far broader social scientific problem' (Guyer, 1987: 47).

However, rather than answering questions, this approach raises them. A historical approach is an important method of contextualising a problem and raising questions which need to be asked. Rather than an agenda for research being introduced (or imposed), questions are raised and directions for seeking answers suggested. For example, Bryceson (1985, 1993) reconstructs and analyses the development of the food supply in Dar es Salaam from government documents, researcher accounts and other sources, through periods of crisis and balance from the beginning of the last century. She concludes that it is simplistic to promote the balancing of Tanzania's external trade and finance and the lifting of government controls as solutions to the country's food deficit problem. There are long-term issues which must be considered, not least of which are the socio-political impacts of such measures. Straightforward economic measures, she argues, are steeped in ideological debates concerning the role of the state in the economy, while attention should be paid to the impacts of such measures on the stability of the economy and, ultimately, on the people. She suggests that a series of policies should be implemented which are aimed at curbing rural–urban migration, at firing the initiative of the trader and producer and at encouraging a sense of professionalism in the bureaucrat. However, she omits what the details of these policies are to be, how they are to achieve their objectives, and what role should be played by government, traders, producers and consumers. She suggests that changes to one rural–urban flow may have implications for others and that reducing rural-to-urban migration will constrain urban growth as a positive approach to food supply, but does not describe how this will work. Indeed, other authors may take issue on the basis that migration to cities increases the food markets and there is some evidence that this may have a positive effect on food production. This illustrates the need for more integrated approaches which allow for comparisons.

It appears that each of the above traditions tends to concentrate on one or two areas of research at the expense of the rest. The first three approaches provide us with detailed analyses of different aspects of the food supply chain. An approach concentrating on quantitative economic data has inherent problems involved in data collection and interpretation, as it tends to focus on prices, volumes and values of formal market transfers. An ethno-geographic approach tends to focus on representations of the markets and fairs where the food is bought and sold, with an emphasis on the cultural importance of markets to a variety of interest groups and institutions. An approach concentrating on the impact of the food marketing system on the urban population's welfare can over-emphasise the role of the state at the expense of the already efficient marketing system. This approach can identify institutional defects in the access to the supply of food of the urban poor. The historical approach provides a good starting point, contextualising the problem of food supply by setting in place the cultural and institutional legacy of the society. Each of these four approaches may reproduce a valuable view of an urban food supply system. However, any picture of urban food supply which is painted in only one of these dimensions will ultimately lack the overall depth which can be achieved by using a far broader conceptual framework.

Spatial market approach

Meanwhile, a further approach can be distinguished which concentrates on spatial market networks. It has had a limited impact outside geography, but within the discipline it is widely regarded as being of great influence. The spatial approach is applied mainly to rural marketing systems, focusing on attempts to explain the phenomenon of periodic marketing. There are a limited number of studies which have examined urban marketing. Barrett (1986) distinguished two explanations of market development within a geographical approach to market systems in developing countries: central place theory and the diffusionist approach. The central place explanation is based on Skinner's (1964/65) study of trade and marketing structures in rural China. Skinner identified systems of market towns connected through flows of goods and services, and movements of traders and consumers. These spatial linkages resulted in hierarchical systems of markets with overlapping hexagonal hinterlands. The explanation of this hierarchical system is that trading takes place between small peasant centres and these evolve at varying rates depending on their location in relation to their hinterland.

The diffusionist approach is based on the work of Hagerstrand (1952) on historical records of innovation diffusion in central Sweden. This follows a three-stage process: first, multiple centres of adoption emerge; second, new centres grow up and compete with the initial adoption centres; and finally, additional increments are not feasible as all centres have adopted the innovation and a stage of saturation is reached. An innovation is more likely to be adopted close to an early adoption centre and the probability of new centres emerging reduces with distance from it.

The first theory assumes the importance of the evolution from peasant-to-peasant trading to rural-to-urban trading. The second sees a process of 'top-down' penetration of the market system. Barrett (1986) suggests that each theory may have greater power of explanation depending on the local conditions. For example, she argues that in studies where the diffusionist explanation has been successfully applied, the growth of the market system has taken place relatively recently, and has occurred in conjunction with many other changes such as improved socio-economic conditions and transport provision. In areas where central place theory has been successfully applied, markets are already in place, and the adoption of the market system moves outward from established centres, forming hierarchical relationships between centres of differing importance in the marketing system. This process tends to strengthen the position of the established markets and results in a hierarchical pattern of rural-to-urban trading. Further dimensions exist within most market systems, however, and these must be examined in order to explain, rather than describe, the development of market systems.

Ideally, the study of an urban food supply system should consider the efficiency of supply to all sectors of the population. Put slightly differently, the ability of all sectors of the population to get access to food supply should be assessed and the real cost of that access investigated. It is important that such research addresses itself to the role of the various actors and structures involved in the food supply chain, such as the government, city or state marketing authority, the national infrastructure, the migrant traders, the producers, the consumers and those involved in all intermediate stages of buying and selling. The relations between these institutions require examination. An assessment is also required of the effect of the relative power or influence of these institutions in the system. Clearly such investigations must incorporate the formal and informal aspects of the food supply system, since estimates of the informal sector's role in food supply, even in countries with heavily managed food markets, are relatively significant. Sporrek (1985) estimates that the informal sector accounts for 50 per cent of food marketing in Dar es Salaam, while other estimates suggest similar figures for cities around the developing world. This requires an economic analysis of the price and volume data held by formal marketing agencies, as well as attempts at data capture within the often elusive informal sector. More recent geographical research has integrated a number of concerns, the spatial dimension providing limited explanatory power to analysis of food markets. For example, in calling for an overall framework for the study of urban food distribution systems in Africa and Asia, Drakakis-Smith suggested that geographers are now well placed to investigate the broad issues involved. He suggested this is because they 'must focus not just on the commodity itself, but also on the economic, social, cultural and political factors which affect the system' (1991: 51).

Drakakis-Smith (1990) identified three main elements of the food supply system:

- food-producing areas (domestic rural and urban, international);
- food-marketing networks;
- urban consumption centres.

Figure 2.2 illustrates the links between different actors in urban food supply networks. These networks operate across the urban–rural divide at various points. In particular, food-producing areas and food marketing networks can be located in both urban and rural areas. As rapid urban growth takes place, so the network of a city's food supply system tends to spread geographically outwards in search of more territory from which to source food for the growing urban market. In addition, the majority of developing countries have a significant and growing dependence on food imports. Most of these countries are net food importers. Much of this imported food goes straight into urban markets. Thus the rapid growth of the urban population results in growing dependence on food imports which threatens not only urban food security but also the overall food security of some of the world's poorest countries. In addition to the

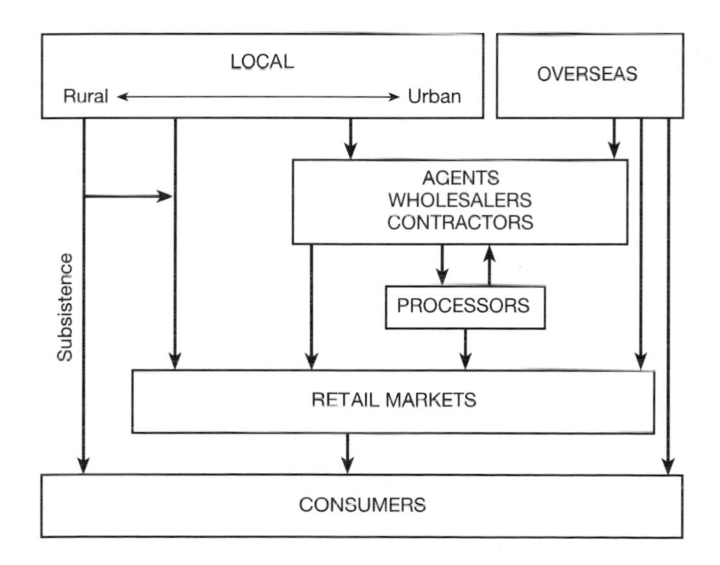

Figure 2.2 ***The urban food supply system.***

Source: adapted from Lynch (1992) and Drakakis-Smith (1990).

issue of food security, the increasing dependence on food imports into cities represents a divergence from the role envisaged for rural areas in Rostow's (1960) model of economic growth (discussed in Chapter 1 above). In this model, rural agricultural production supplies urbanising areas with food and this in turn develops demand for manufactured commodities and agricultural inputs which can be supplied by the developing urban-based industries. Where cities in developing countries increasingly rely on imported food, they are less reliant on their rural hinterlands, the linkages are diminished and this may affect the ability of both to develop. Urban food supply is therefore an issue of importance not only for the sustainable development of cities but also for the national food security of many developing countries.

Food-producing areas

In most developing countries agriculture forms the dominant productive economic sector. Most rural dwellers are directly dependent on agriculture for their survival. The very rapid growth rates experienced by many cities in developing countries mean that in many cases more than half the urban population were born into rural agricultural households before migrating to the city. Some authors describe this trend as the 'peasantisation' of the city, as rural dwellers bring their rural values, skills and lifestyles to the city (Roberts, 1995).

In archaeological literature attention is given to the role of food production as this was one of the key economic activities in ancient civilizations. There is evidence, for example, of food commodities in archaeological sites in the Indus Valley that have been dated as older than 2,000 BC with extensive irrigation, granaries and other workings to serve cities with populations of over 35,000 (Mougeot, 1999). Such developments facilitated not only the production of food but also the processing, such as different ways of cooking, preparing and milling.

Archaeological evidence suggests that the reasons for city formation may be related to the formation of what we understand to be countryside, in that the development of cities is often related to the development from hunter-gatherer systems to settled agricultural systems. The interdependent links between city and rural area are a key factor in the early formation of the cities and in the development of sedentary agriculture.

There are two main theories that relate to the formation of cities. The first is that the collection of a large number of people in a settlement compound lessens the threat of internal conflicts between members of the community. This can be seen as a conflict resolution motivation. However, such spatial concentration of population can attract different threats from outside the community. An alternative theory argues that the collection of a large number of people facilitates the development of large and complex subsystems, in particular the construction and management of projects such as irrigation, fortification or granaries. For example, some irrigation canals in ancient Mesopotamia were 40 kilometres long. Much of the evidence suggests that from these early beginnings there are few examples of cities which did not quickly become seats of power for ruling elites. Thus cities were the centres of decision making, social control and integration. There is a debate about whether the role of cities is positive, sometimes described as 'generative' (Gilbert and Gugler, 1992), providing the initiation for the development of the region (city and countryside). Alternatively, the city may play an extractive role of organising the rural peasant class in order to take surplus from the countryside. This debate continues today, focusing in particular on whether government policy has been urban biased or whether it has genuinely attempted to develop the countryside. In Rostow's (1960) stages of economic growth model the role of cities appears to be conceived as both generative and extractive. They extract surplus which is invested in urban-based development and industrialisation, and which, in turn, generates economic activity which provides inputs for rural production and demand for rural produce.

More recently, economic activities other than food production have become relatively more important, with cities acting as a focus for manufacturing, trading, processing and other services, as well as food supply. This development of other industries, however, has not entirely diminished the political importance of food supply to urban centres. More advanced economies have spread their risks, sourcing food not only from their rural hinterlands but also from producing countries across the globe. This is possible because the size of the markets and the economies of scale of industrial food manufacturers and supermarkets can bear such distant sourcing. In developing countries, food production is still a largely rural activity and most cities rely more heavily on their domestic production of food. Food purchasing for consumption, on the other hand, is still a largely urban activity. Therefore food supply is still a key aspect of rural–urban relations in the developing world. In part, this seems to have come about as a result of the importance of cities as markets for developing rural agricultural production economies. What happens in the city food markets is therefore very important to the rural food producers. In less developed countries, food production for urban markets is often seen as a way of integrating remote rural areas into the economy of the rest of the country and a way of encouraging subsistence producers to enter the economy and maximise the productivity and therefore the food security of the economy. It could be argued that the persistence of subsistence economies is also an indication of the failure of the markets. The market has failed to provide sufficient incentives and security to encourage commercial peasant production which would increase rural–urban economic linkages. Von Braun and Kennedy (1994) argue that 440 million farmers in the developing world continue to farm for subsistence on 50 per cent of land in most low-income countries. This is despite the need for change to respond to resource constraints, land degradation and rapid urbanisation. Commercialisation of agriculture, von Braun *et al.* (1993) argue, will result in increased food production, improved urban and rural household food security, increased rural and urban employment and an increased circulation of capital between rural and urban areas.

As cities have grown, so urban markets have become more important as outlets for rural produce in the rural population's survival strategies. However, cities have also begun importing from foreign sources. The corollary is that processed goods sourced from urban industries have been important in rural areas.

Food is still a key to the survival of both rural and urban populations. Even in some countries that have benefited to some extent from development, the role of markets can still be relatively important. Figure 2.3 illustrates this with a picture of a market in Nebeul, Tunisia. It is an important wholesale market, as it is close to the fruit-growing areas of Cap Bon, but it is also an important urban market, where rural producers bring fresh produce to sell to both wholesalers and consumers. In the background a farmer or rural-based trader sells his produce off the back of his lorry, while in front of the lorry some retailers sell to the consumers and other smaller retailers (in the foreground).

One of the consequences of the selectivity of rural–urban migration (see Chapter 4 below), is that those arriving in cities may be affected

Figure 2.3 ***Fresh food trade in Nabeul market, Tunisia.***

by what Holm (1994) calls the 'urban reproduction squeeze'. Under these conditions, few people are able to maintain their reproductive needs with any one economic activity, so they are forced to engage in a range of activities including subsistence and market-based activities. Urban residents experiencing the reproduction squeeze respond by seeking a wide range of livelihood options. Bryceson (1993) reports that migrants in intermediate towns in Tanzania grow a part of their household staple food needs. In contrast, around a third of respondents received food supplies from extra-household sources, most of which were members of the extended family. However, Holm (1994) reports, in a study of intermediate Tanzanian towns, that where most of the urban cultivating households surveyed were male-headed, many of the female-headed households that migrated to the city did so in protest at the patriarchal system in the rural areas. This seemed to result in their not being able to rely on food sent to the intermediate towns surveyed, with 12.8 per cent sourcing from their rural home as compared with 23 per cent in the case of male-headed households (Holm, 1994). This suggests that the female-headed migrant households are far more dependent on urban-based livelihood opportunities than their male-headed counterparts.

A final type of food-producing area supplying urban areas is that where food is cultivated in the city itself. As highlighted earlier, there is archaeological evidence that food was produced in cities in the pre-colonial period. There is also evidence of food being produced in cities during the colonial period. However, towards the end of the twentieth century, research was beginning to suggest that urban agriculture was experiencing something of a renaissance. This is discussed later in this chapter.

Food marketing networks

A key element in the urban food supply system is the market. Kaynak (1986) argues that in the past the trading functions (in this case he is referring to the general meaning of buying and selling) in less developed countries were ignored in favour of extractive or manufacturing activities and attempts at capital formation. Paddison *et al.* (1990) suggest that this is due to the assumption among economists that investment in the productive sectors of the economy would encourage the development of consumption. Kaynak (1986) argues that this has resulted in neglect of marketing activities for three main reasons. First, he argues that 'traders are widely regarded as "parasites" and as such, trade has a low esteem as a profession'. Second, in many developing countries, trade, and especially wholesale trade, can be dominated by ethnic minorities. Finally, because of these negative attitudes towards trade, 'government policy makers tend to underestimate its contribution to the national economy. In most cases trade accounts for some 20 per cent of the GNP of these countries' (Kaynak, 1986: 17). These three arguments may be contested, but what is clear is that trading in many developing countries is a highly adapted, dynamic sector of the economy, and that its contribution is often underestimated or taken for granted by policy makers and governments.

By describing trading in developing countries as the economic bridge between production and consumption, Kaynak argues that marketing stimulates development rather than being dependent on it, and improvements in marketing 'in any economic system can aid in economic development by leading to a more efficient use of existing productive resources' (Kaynak, 1986: 17). Maluf (1998) goes further, arguing that there is evidence in Latin America of direct links between food systems and economic development, such that there

Figure 2.4 ***Mgeta rural wholesale market in Tanzania.***

is a public requirement to regulate food security in order to provide the multiplier effects that benefit the whole economy. This means not state involvement in ways that have been unsuccessful in the past, but a coordination of state, private and community interests to ensure equitable access to food through increased domestic markets and the stimulation of domestic food production. She recommends 'rebuilding institutional mechanisms in order to achieve the public regulation of economic activities (with food security among the strategic goals) and to improve the synergistic interaction between the state and the society' (Maluf, 1998: 171).

Kaynak (1981) argues that there is a progression in the development of food retailing and distributions systems. He suggests that the urban food marketing systems of developing countries are characterised by a large number of small retailers with a few large central markets. Figures 2.4–2.6 illustrate the three key locations and events in Tanzania that form gateways for rural traders and producers, giving access to urban markets. In Figure 2.4 the Mgeta rural market has sufficient rural food production in its hinterland for export out of the district to distant urban markets to merit the construction and management of a small wholesale market which acts as a focal point for traders and farmers. Farmers carry their crops by headload to the market building, where traders bulk up their loads to make the rental of a lorry commercially viable. Figure 2.5 shows Soni periodic market, where the trade is insufficient to merit the construction of a market building. The response has been what is described as temporal arbitrage, where farmers and traders meet on a regular basis, often weekly, sometimes more frequently. Traders often tour such markets, which often take place on sequential days in order to facilitate this. Periodic markets in Tanzania have a clear gender division of labour. The women usually sell small quantities of produce in piles and often to other women, while the wholesale traders and the producers selling to traders are mainly men. Figure 2.6, the final image in this set, illustrates the Kariakoo market building in central Dar es Salaam, constructed to house the city's wholesale market for fresh fruit and vegetables. It was quickly found to be too small and much wholesaling takes place in peripheral markets close to the main roads running into the city.

At the time Kaynak was writing, in the early 1980s, government was also an important player in most food markets, in particular those for staple foods. Since then, however, the wave of structural adjustment policies which were introduced in most developing countries has

Figure 2.5 ***Soni rural periodic wholesale and retail market, Tanzania.***

Figure 2.6 ***Kariakoo market building, Dar es Salaam, housing the city's main wholesale market.***

swept away most state-managed marketing of food. This is in contrast to the dominance in the developed economies of large supermarkets and hypermarkets. This idea of progression appears to be more marked in intermediate or transition economies, where there is evidence that large outlets are in the ascendancy (Findlay *et al.*, 1990). Kaynak (1981), for example, explores the role of supermarkets in Turkey, and Bromley's (1998) urban retailing studies found supermarkets to be important to urban dwellers in Latin American cities. The significance of this, according to Kaynak (1981), is that these market structures are highly inefficient, leading to high-cost marketing systems. This cost is passed on to the consumer and leads to poor price incentives for the rural farmer. Thus the market signal is very weak by the time it reaches the rural producer. Because of the long supply chain, the range of factors affecting the price for the consumer means that the demand changes are unlikely to be communicated to the producer in any meaningful way through price variations, since so many factors other than consumer demand influence the farm-gate price. It may also be affected by the fact that urban consumers are also producing their own food.

However, Kaynak (1981) is also at pains to explain that a modernistic progression of market structure is unlikely to take place. He specifically mentions Rostow (see Chapter 1 above) as influential on the ideas of Western marketing scholars in their suggestion of a simplistic modernisation model of retailing development.

> What is most troublesome with Rostow's framework, in terms of the contribution of marketing to economic development, is the premise that development processes can somehow be hurried along if institutions from developed economies are transplanted into developing economies without due attention to the varying environmental conditions prevailing in the recipient country.
>
> (Kaynak, 1981: 81)

He suggests that much of the force which brought about the transition in retailing structure in developed economies was that exerted by the demand of the consumer. It is important, he argues, to be aware of the environmental circumstances of the food retail sector. For example, in some countries food consumers are comfortable spending more time shopping than in others. Some can afford to buy large amounts on infrequent shopping trips while others, due to financial and storage constraints, may have to shop a little each day. Thus the marketing and distribution chains linking producer and consumer are many and short, whereas those in the developed world are dominated by a few retail outlets at the end of long marketing chains. A consequence of this is that rural–urban food flows appear to become increasingly dominated by larger companies. This is partly a response to consumer demand and partly a consequence of changing economic and social circumstances. The result is also that these different retail structures offer contrasting services, as illustrated in Table 2.2.

Table 2.2 ***Features of developed and developing retail markets***

Feature	*Developed market*	*Developing market*
Credit	Not available	Large amount
Frequency	Large shop weekly	Small amounts daily
Preparation	Pre-prepared food	Fresh food for home preparation
Role of shopping	Culturally and socially important	Economic necessity – a chore
Service	Self-service	Retailer served

Source: adapted from Kaynak (1981).

An ancillary benefit of food retailing is the contribution of these markets to the local economy. As much of the marketing of food crops in developing countries is informal this is often difficult to quantify. Wilhelm (1997) estimates that government revenues from markets can account for 10 per cent or less of local revenue. The reason for such a low level seems to relate to a number of issues, such as the low level of fees, but also poor collection rates; for example, Wilhelm (1997) estimates that approximately 30 per cent of collected sums are banked. Low levels of pay lead to poor incentives to collect and various types of corruption. In Dar es Salaam, this author met market officials sent to illegal markets to attempt to collect duties and commission fees from traders. They are usually alone and travel by bus or cycle. They find their work difficult when the city provides no facilities. The reason for illegal markets is generally that the existing legal provision is overcrowded and poorly maintained. The officials have to collect what they can and usually end up haggling with traders to achieve any collection at all. In addition, markets provide for distribution of other commodities: for example, rural markets attract sellers of processed food and manufactured goods.

On the other hand, Wilhelm (1997) argues that poorly run market sites can add to a city's costs. For example, lack of facilities and constraints on distribution can add to the prices paid by consumers. Markets which burst at the seams onto adjacent roads can cause major traffic disruption. Poor conditions in markets can lead to poor sanitation and hygiene, increasing the health hazards due to food handling. However, Polly Hill (1963) anticipated such arguments in her response to Bohannan and Dalton (1962) when she argued forcefully that predictions of the end of market in Africa (i.e. Bohannan and Dalton, 1962, and the contributors to their edited volume) were premature. She argued that the apparent chaos of African markets is usually based on Eurocentric ideas of how an ideal market should operate. This view is subjective rather than being based on empirical research into how African markets actually work. She suggested that these 'uncomfortable and inconvenient places' (1963: 444), rather than putting off researchers, should encourage them to develop new techniques, the better to understand how the markets work. Lado argues that food marketing systems are not only important for the supply of food to the city but also 'transmit the impact of macroeconomic policy on to farmers' incentives and therefore onto production and distribution' (1987: 373).

Box 2.1

Commercially successful urban agriculture in Jakarta, Indonesia

In the early 1970s a young man called Bob Sadino saw a market for special food products in Jakarta, Indonesia. He began importing chicks from the Netherlands, raising them at his parents' home and selling the eggs to neighbours. Within four years he had developed his business, supplying dressed chicken to hotels and chicks to other poultry farmers, establishing a retail outlet in his family home, and buying a meat-processing plant where he processed chicken and other meats, while continuing his door-to-door business. He now has a multimillion-dollar urban agriculture business called Kem Chicks, which employs 800 people, contracts outgrowers and exports high-value speciality foods.

This case study is an interesting illustration of a number of issues which have arisen in this chapter. The development of Kem Chicks illustrates the way in which a small-scale informal-sector activity links with international business, in that Sadino imported his chicks into his parents' house and gradually moved towards supplying hotels and other poultry farmers. The case study illustrates the possibility of crossing from the informal sector into the formal as Sadino's business takes a name, employs hundreds and develops, expands and invests. Finally the case study illustrates one of the forms of urban-based agriculture, small-scale livestock rearing, that benefits from excellent links to its market through its urban location.

Adapted from Urban Agriculture Network (1996).

In other words, the organisation of the marketing system is crucial to any strategies for improving the peasant production of food. The significance of food flows to effective urban–rural linkages is illustrated by Boxes 2.1 and 2.2.

Urban and peri-urban agriculture

During the 1980s there developed an increasing interest in the phenomenon referred to in recent years as urban agriculture. Initially it was thought that this was rare and largely discouraged by city authorities and governments as being counter to urban modernism. However, over time, as a coherent literature has developed, it has become clear that urban – and peri-urban – agriculture has been around for some considerable time.

Table 2.3 Farming systems common to urban areas

Farming system	*Location*	*Technique*	*Product*
Aquaculture	Ponds, streams, estuaries, sewage, lagoons, wetlands	Cages	Fish and seafood, vegetables, seaweed and fodder
Horticulture	Homesites, parks, rights-of-way, roof-tops, wetlands	Hydroponics, greenhouses, containers	Vegetables, fruit, compost
Livestock	Rights-of-way, hillsides, peri-urban, open spaces	Zero-grazing, coops	Milk and eggs, meat, manure, hides and fur
Agro-forestry	Street trees, homesites, greenbelts, wetlands, hedgerows	Orchards, steep slopes, vineyards, forest parks	Fuel, fruit and nuts, compost, building materials
Other	Sheds, rights-of-way, roof-tops, urban forests	Ornamental horticulture, beehives/cages	Houseplants, medicine, beverages, herbs, flowers, insecticides

Note: Zero-grazing is a method of animal husbandry that involves bringing fodder to the animal to prevent damage to vegetation and control feeding.

Source: adapted from Urban Agriculture Network (1996).

A comprehensive review of forms of this cultivation from all regions of the world provides a typology of farming systems identifiable in urban areas. It includes aquaculture, horticulture, livestock rearing and agro-forestry (see Table 2.3).

Bryceson (1993) provides documentary evidence from as early as 1939 of the importance of urban agriculture during the Second World War and after. However, it was only in the mid-1980s with the UN University's Food Energy Nexus research programme that researchers became more serious about examining agricultural cultivation within city boundaries. Since then a significant literature has developed around a central debate on whether urban agriculture is a good thing. This has been summarised by Lynch (2002) and is illustrated in Table 2.4. However, one problem with this summary is the assumption that urban agriculture is a homogenous activity with universally similar impacts. A second problem with it is that it is largely viewed from the perspective of the city's population. This means that urban agriculture is generally seen as beneficial, because it provides food and employment and can even provide environmental benefits. However, there is little consideration of the impacts on the rural–urban food supply. This led Lynch, Binns and Olofin (2001) to argue that urban agriculture should be considered as a part of the more general urban food supply system. This

Box 2.2

Successful cooperative urban agriculture in Dakar, Senegal

In the community of Pikine in Dakar, Senegal, a cooperative of small entrepreneurs has succeeded in farming in a wetland area of tribal land which is unsuitable for building. The cultivation is mostly done by men, who grow vegetables under trees and raise livestock, mainly for the market. The marketing is done by women. Both men and women process and sell related products such as dried fish, tanned leather and handicrafts made from palm frond. In addition, the marshier parts are leased to rural itinerant rice farmers and the rent used for common projects. The farmers collect waste from households, markets and animals – and occasionally from sewage pipes – to fertilise and irrigate the soil. The cooperative operates under the leadership of an elected president, who is also the tribal chief. The success seems to derive from strong organisational structure and the integration of marketing, processing and land management.

This case study illustrates a successful form of urban cultivation which is focused around a cooperative. The cooperation of the members and structure of their organisation delivers a measure of stability and trust that could otherwise be absent. The farmers' activities benefit from the supply of sewage and waste from the area of Dakar in which they are based and from the proximity of their market. They are also able to make use of urban land that is otherwise unusable.

Adapted from Urban Agriculture Network (1996).

Table 2.4 ***Summary table of urban agriculture research***

Advantages	*Concerns*
Vital or useful supplement to food procurement strategies (Rakodi, 1988)	Conflict over water supply, particularly in arid or semi-arid areas (Mvena *et al.*, 1991)
Various environmental benefits (Lynch, 1995)	Health concerns, particularly from use of contaminated wastes (Lewcock, 1995)
Employment creation for the jobless (Sawio, 1994)	Conflicting urban land issues (Lynch *et al.*, forthcoming)
Providing a survival strategy for low-income urban residents (Lee-Smith and Memon, 1994)	Focus on urban cultivation activities rather than broader urban management issues (Rakodi, 1988)
Making use of urban wastes (Egziabher, 1994)	Urban agriculture can benefit only the wealthier city dwellers in some cases (Smith, 1998)

Source: adapted from Lynch (2002).

Box 2.3

Air pollution and urban and peri-urban agriculture in South Asia

Some writers have put forward the theory that as urban populations increase in affluence, there is evidence to suggest that their urban environment declines in quality (McGranahan and Satterthwaite, 2002). In general, Afsar (1999) found the risk of health hazards from air pollution to be greater in urban Bangladesh than in its rural areas (see also Chapter 3 below). Poole *et al.* (2002) have pointed out that the decline in the quality of the urban environment can take a double toll if food is also produced there. Air pollution can affect food crops by reducing yield and reducing the nutritional quality of the crop; for example, sulphur dioxide is known to cause significant crop yield losses close to urban industries, while high levels of ozone affect crop production on the edge of cities (see for example Figure 2.7).

Tropical countries appear to be particularly vulnerable to air pollution effects for two reasons. First, in countries with greater levels of solar intensity, crop damage can be more severe because the sunlight accelerates the chemical reactions that produce ozone in the lower atmosphere. Second, tropical crop cultivars can be more sensitive to adverse effects from ozone. Phytotoxic gases are reported to result in yield reductions of up to 40 per cent on the outskirts of Lahore, Pakistan, and in Varanasi, India.

Figure 2.7 ***Urban food systems located next to industrial activities near Delhi.***
Source: Nigel Poole, Imperial College London.

Sources of pollutants include metals and chemicals industries, textiles, coal, transport and the production of a range of additional industries such as plastics, carpet making, construction and pottery. The crops can be affected as they are growing – particularly, for example, if they are grown on roadside or railway-side plots or close to industrial plants. In addition, since many foods are sold by the side of roads in order to take advantage of the passing trade, the air pollution from traffic can further affect the quality of the produce. Field data collection found levels of cadmium and zinc in a range of retail sites around Varanasi, India, with zinc being high during summer and cadmium being high all year round. In each case the amounts exceeded permitted levels. In a survey of market agents there was found to be a high proportion of respondents who judged the quality of their produce by its freshness, variety and colour. This latter is important as the discolouring effects of contamination by air pollutants varies considerably from one product to another. For example, cauliflower, which is one of the vegetables most likely to trap heavy metals, is also least likely to show this by discolouration.

The improvement of food safety in India would require initiatives at the national level involving central and state government, at the meso-level involving institutions in the marketing chain, from producers to consumers, and at the micro-level involving the individual actors in the production, marketing and consumption of the produce. This will require the involvement of public, private and scientific bodies working in partnership for improved environmental health.

Adapted from Poole *et al.* (2002).

facilitates consideration of the relative benefits of issues such as the city providing a market for rural production and the question of whether cities should use scarce water supplies for urban cultivation if nearby rural areas can provide food with less pressure on water resources. It would also facilitate the consideration of the relationship between environmental management and urban food cultivation. Box 2.3 illustrates some of the issues relating to urban food cultivation and urban-related environmental problems with the example of air pollution in India.

Concerns voiced by farmers in developing countries are not dissimilar to those voiced by peri-urban farmers in the developed world. Bryant and Johnston (1992) carried out a survey in the US of farming in what they called the 'city's countryside' and found that key among concerns of farmers were changes in the local community, there being too much traffic, the problems of trespassing and vandalism, and complaints from non-farm neighbours.

Conclusion

This chapter has considered a range of approaches to the topic of food flows to the city in the developing world. It has reviewed a wide range of approaches that have been applied in what has been described as a sparse literature – at least in relation to other services in urban areas. Concern has been expressed by some researchers that the way in which food flows have previously been researched has been partial. Latterly interdisciplinary approaches have been emphasised, which reflect the wide-ranging set of connections between food supply systems and other aspects of the economy and society. It is argued that this broader approach should also take into consideration alternative sources of food supply such as food imports and urban production in order to assess the impacts on both urban and rural areas and the linkages between them.

Discussion questions

- Review the main approaches to the study of food supply for the cities of the developing world.
- Consider the foods and drinks that you have consumed within the past 24 hours. What kinds of production and processing activities have prepared it for you and have these taken place in urban or rural areas?
- In what ways can cities contribute to food systems in the rural areas of developing countries?

Suggested reading

Drakakis-Smith, D. (1991) Urban food distribution in Asia and Africa. *The Geographical Journal* 157(1), March, 51–61.

Heilig, G. (1999) *Can China Feed Itself? A System for Evaluation and Policy Analysis.* International Institute for Applied Systems Analysis, Vienna. Also at http://www.iiasa.ac.at/Research/LUC/ChinaFood/Inde_m.htm [last accessed: 29 November 2003].

Lynch, K. (2002) Urban agriculture. In V. Desai and R. Potter (eds) *Arnold Companion to Development Studies.* Edward Arnold, London. 268–272.

Mortimore, M. (1998) *Roots in the African Dust; Sustaining the Drylands.* Cambridge University Press, Cambridge.

3 Natural flows

Summary

- The development of cities has a variety of impacts both within the city and in rural areas.
- As cities become more affluent they minimise local impacts and may transfer their environmental burden to more distant locations including rural areas.
- Environmental agendas are socially constructed and vary from one society to another.
- Cities and rural areas can provide environmental benefits or can adversely affect the environment.
- One area in which environmental management is particularly challenging is the rural–urban interface.

Introduction

This chapter considers the environmental linkages between urban and rural areas, in terms of both the positive element of this relationship, resources, and the negative element, the constraints and hazards. When considering natural resources, there are two key aspects to urban–rural interactions. These focus first on the issue of resources produced by rural areas and wanted in urban areas. Second, these relate to the way in which urban environmental management can impact on the rural areas. A cursory examination of these relations would suggest that the cities obtain the best of the arrangement, benefiting from the resources and labour of the rural areas and disposing of pollutants to the rural areas. Hardoy *et al.* list the following main ways in which the inhabitants and the environment are affected by the development of a city:

> - The expansion of the built-up area and the transformations this brings – for instance land surfaces are reshaped, valleys and swamps filled, large volumes of clay, sand and gravel and crushed rock are extracted and moved, water sources tapped and rivers and streams channelled.
> - The demand from city-based enterprises, households and institutions for the products of forests, rangelands, farmlands, watersheds or aquatic ecosystems that are outside its boundaries.
> - The solid, liquid and air-borne wastes generated within the city and transferred to the region around it which have environmental impacts, especially on water bodies where liquid wastes are disposed of without adequate treatment and on land sites where solid wastes are dumped without the measures to limit their environmental impacts.
>
> (2001: 172–173)

Cities require high absolute volumes of resources (especially water, energy resources, land, food and raw materials). For example, Rees (1997), the person credited with the development of the ecological footprint approach (see Box 3.1 and later discussion), found that Vancouver, Canada, required the productive output of 180 times its own land area. O'Meara (1999) argues that available data at a global scale suggest that 78 per cent of industrial carbon emissions and 76 per cent of industrial wood use occur in cities. Some 60 per cent of the planet's water resources goes directly to cities or is used to irrigate food crops and by industry. This latter is referred to by Allen (2001) as 'virtual water' – it is not directly consumed by urban residents, but is ultimately consumed by them in the form of foodstuffs or consumer goods that have been produced using this

Box 3.1

Ecological footprints: the global and inter-generational impacts of urban lifestyles

A recent report by the Organisation for Economic Cooperation and Development concluded that large and wealthy cities focus the demand for food, fuel and raw materials. This demand is often met by importing from increasingly distant source areas. The result is that high living standards are enjoyed by affluent urban dwellers, while natural resources and landscapes around them are preserved. However, the result is a system that can transfer costs to people and environments in other regions or countries. Urban lifestyles can also transfer the impacts and costs of their consumption into the future. For example, hazardous wastes are a long-term threat to human health and ecosystems. The economic costs of unsustainable consumption patterns – forest or fish-stock depletion and biodiversity loss – also affect the future. While wealthy cities may cut emissions of the pollutants that affect their local environments, carbon dioxide emissions continue to rise because of the dominance of fossil fuel-burning private motor transport. The consequences of carbon dioxide emissions may affect climate at a global scale and will be felt by future generations.

The OECD argues:

> The calculation of cities' ecological footprints, developed by William Rees (1992) . . . makes evident the large land area on whose production the inhabitants and businesses of any city depend for food, other renewable resources and the absorption of carbon to compensate for the carbon dioxide emitted from fossil fuel use.
>
> (OECD, 1996: 45)

However, most urban centres in developing countries have very small ecological footprints. The ecological costs of urban-based activities are due not to their being urban as such, but to the lifestyles and consumption patterns of their consumers and the resource use and waste generation patterns of their enterprises and institutions. As discussed above, studies applying Rees's (1992) methods, analysing the ecological footprint of cities, have shown that large and wealthy cities in developed countries do have large 'ecological footprints'. However, there is also evidence that high-income households in rural or suburban areas often have larger ecological footprints than comparable households in urban areas, largely because of greater car use and higher levels of energy use for space heating or cooling (Rees, 1997; Wackernagel, 1998).

The OECD concludes that in order to make urban development sustainable, the following approaches are required:

- improving the health and safety of urban environments;
- reducing the environmental burdens that urban-based production and consumption transfer beyond their boundaries, for both present and future generations; and
- (where appropriate) building upon the synergies (and avoiding the conflicts) between poverty reduction and environmental improvement.

Adapted from OECD (1996: 94).

water. The greater the population and the more affluent they are, the greater the demand for resources. However, the literature suggests that this relationship is not necessarily a direct one. There are researchers who have argued that cities also have beneficial environmental effects (see later discussion), though the net environmental effects may be more important in the population–environment relationship.

The environmental burden

McGranahan and Satterthwaite (2002) illustrate the importance of the issue of scale. They point out that the wealthiest cities score highly when their sustainability is measured against the extent to which they meet the environmental needs of their inhabitants. Their analysis suggests that a 'transfer of environmental burdens' takes place where wealthier cities meet the needs of their inhabitants by transferring their environmental burden elsewhere, by sourcing raw materials from increasingly great distances and limiting pollution to reduce local effects but perhaps with wider implications. McGranahan and Satterthwaite (2002) argue that cities that fail to meet the needs of their inhabitants do so because they are less able to transfer environmental burdens, which therefore land on their own populations. For example, in cities where the transfer of the environmental burden is low, the risk from pollution in the city is likely to be high. Affluent cities tend to have regulated against local pollution effects. This is also a feature of intra-urban relationships where wealthy urban residents have the highest transfer, such that their environmental concerns are minimal, while low-income residential areas tend to have high environmental risks, such as from landslips, pollution and disease. Thus, in intermediate economies wealthy urban residents are able to afford to reduce their risk by transferring the environmental burden, for example by living further from hazardous or polluting industrial areas. In some tourist resorts in very arid developing areas, water is supplied to hotels for watering gardens and golf courses, while it may be in short supply for local low-income communities. The concept of the transfer of the environmental burden is illustrated in the stylised diagram in Figure 3.1. The relationships illustrated here show that localised environmental burdens decline with increasing urban affluence; while city-region burdens, such as air pollution, increase initially with rising industrial activity, then, as affluence spreads, the city develops

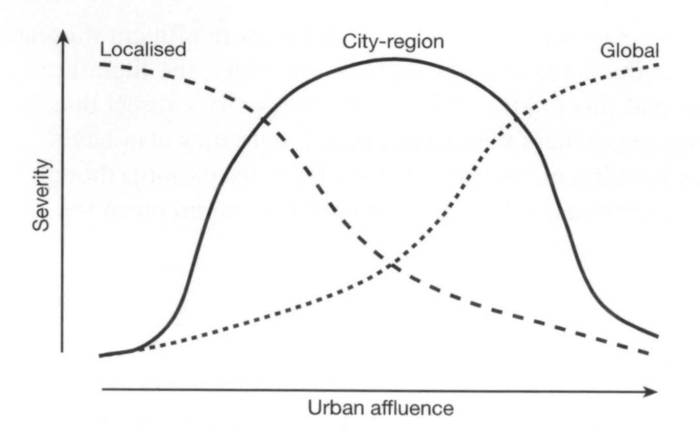

Figure 3.1 ***Stylised illustration of the link between affluence and environmental burdens in urban areas.***

Source: G. McGranahan, P. Jacobi, J. Songson, C. Surjadi and M. Kjellén (2001) *The Citizens at Risk: from Urban Sanitation to Sustainable Cities.* Earthscan, London.

the capacity to control and alleviate such problems. In the meantime activities in a poor city are unlikely to have major global environmental impacts. This capacity, however, increases with affluence.

The problems of water pollution and sanitation decline with increasing affluence, but the effects are immediate and local (see Box 3.1). By contrast, the environmental burdens of carbon emissions are an environmental burden of global concern. The costs of reducing carbon emissions are borne by those taking preventive action, but the environmental burden is borne globally and unevenly according to vulnerability to risk. This means that the greater the wealth of a city then the greater its ability to transfer environmental burdens both spatially, in terms of how far away they may be felt, and temporally, in terms of their implications for future generations. McGranahan and Satterthwaite (2002) also illustrate these relationships in Table 3.1.

Archaeological evidence suggests that the development of cities may give some protection from environmental hazards such as flooding or drought by establishing social and political systems that provide the resources for reconstruction or that establish preventive measures. There is, however, growing evidence that urban living also increases hazards of other kinds. Historically, it is thought that one of the reasons for forming settlements was the pooling of resources in order to reduce vulnerability to environmental hazards such as flooding and

Table 3.1 Urban environmental burdens at different spatial scales

	Local	*City-regional*	*Global*
Air	Indoor air pollution	Ambient air pollutions and acid precipitation	Contribution to carbon emissions
Water	Inadequate household access to water	Pollution of local water bodies	Aggregate water consumption
Waste	Unsafe household and neighbourhood waste handling	Unsafe or ecologically destructive disposal of collected wastes	Aggregate waste generation

Source: McGranahan and Satterthwaite (2002).

drought (Gilbert and Gugler, 1992). However, research on urban health hazards now shows that while infectious diseases decline with increasing urbanisation and wealth, there is evidence to suggest that certain health hazards increase among the poorest urban residents as a result of industrial development (Stephens, 2000). Common urban hazards include pollution (airborne and waterborne), large-scale industrial accidents and terrorist attacks. For example, there is evidence of a link between exposure to air pollution in general and a high rate of asthma among urban children (Davis and Svensgaard, 1987). Stephens reports evidence from Calcutta of a so-called 'double burden' of urban health hazards that come about as a result of increasing urban affluence whereby infectious diseases associated with poverty affect children, while environmentally influenced diseases associated with industrial pollution increase with age (see Figure 3.2). 'Put bluntly, even if the people of Calcutta survive the traditional diseases of urban poverty, they soon feel the impacts of the diseases of an urban environment characterised by "dirty" industrialisation' (Stephens, 2000: 98). The dilemma is that many of the people affected by 'dirty' industrialisation need it for employment and livelihoods. As industrialisation expands, so infectious disease may become less of a problem as wealth and ability to pay for health care increase. There are exceptions to this, in particular HIV/AIDS (see Box 3.2). The wealthier residents are able to protect themselves from such effects, but the urban poor are often vulnerable because of the nature of their work, the location and fabric of their dwellings, and lack of access to preventive and curative medical expertise.

In addition to the direct effects of industrial processing and manufacturing there is also concern about the side-effects of consumption. For example, an increase in car ownership in cities

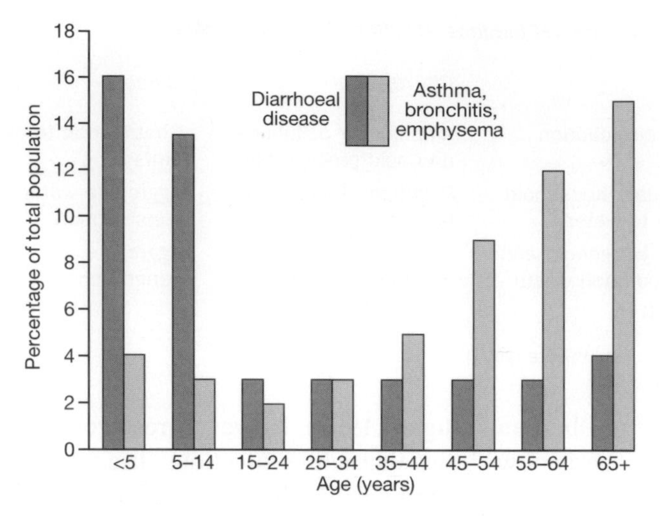

Figure 3.2 ***Double health burdens in Calcutta.***

Source: C. Stephens (2000) Inequalities in urban environments, health and power: reflections on theory and practice. In C. Pugh (ed.) *Sustainable Cities in Developing Countries.* Earthscan, London.

Box 3.2

The urban–rural transmission of HIV/AIDS

Despite the suggestion that increasing industrial development may help to reduce transmission of infectious disease, one disease that is currently affecting many parts of the world is HIV/AIDS. A report in the UK's *Guardian* (12 November 2002) newspaper suggested that truck drivers may be facilitating the transmission of HIV/AIDS in India from city to city and from the city into the countryside. 'A lot of truck drivers frequent prostitutes. We are away from home a lot,' a 21-year-old driver called Kilas is quoted as saying. 'The going rate for sex is 50–100 rupees (70p–£1.40) . . . I earn 10,000 rupees a month so visiting prostitutes is not expensive.' Most drivers are unwilling to use condoms. The result is that an estimated 0.5 million of the country's estimated 4 million HIV-positive people are truck drivers. When drivers return home they may infect their wives. Such are the levels of poverty among the prostitutes that they are willing to risk unprotected sex in return for extra money. This is an illustration of a particularly grim aspect of the interaction between cities and rural areas.

in developing countries will result in very rapid increases in air pollution, with consequent effects on asthma and other illnesses. This is of particular concern because the evidence suggests that socio-economic status plays a role in an individual's ability to avoid the health effects of pollution (Hardoy *et al.*, 2001). In addition, common environmental hazards, such as earthquakes, epidemics or flooding, can affect a larger number of people because of the concentration of population. The literature on the ecological footprint of cities suggests that the environmental impacts of increasingly wealthy cities are displaced to distant locations, whereas low-income cities are unable to do this and the environmental burden falls on their own residents (see Figure 3.1). In terms of rural–urban relationships there are a number of such hazards which involve interaction. McGranahan and Satterthwaite (2000) suggest that it is possible to distinguish between what they describe as 'brown' and 'green' agendas in the initiatives for environmental improvements. These are summarised in Table 3.2. McGranahan and Satterthwaite (2000) list their characteristics in order to highlight the differences between the two, for example the significance of environmental impacts on future generations, impacts on low-income groups and, not least, the effects of city-based consumption on rural resources and ecosystems.

In reality, of course, it is not always possible to distinguish between these two agendas. For example, there may be variations in impacts on low-income groups in different local circumstances. Individuals, whether low-income or elite, make their own trade-offs in terms of their livelihood assets and opportunities. This leads McGranahan and Satterthwaite (2000) to argue for a reconciliation of the brown and green agendas although different cities, neighbourhoods and their linked countryside have different priorities. Indeed they discuss urban agriculture as an example of an activity important to livelihoods of low-income urban residents, which being based on low-income groups' access to land, combines the brown and green agenda perspectives on the advantages of increased local production (see Chapter 2 above). McGranahan and Satterthwaite (2000) give other examples of where the agendas have been combined: these include public transport provision, solid waste management, water supply and sanitation. However, much of this discussion still focuses on the *within*-city agenda. There is, for example, a need to address these agendas and combine them as regards the impacts of urban development on rural people and environments (Main, 1995).

Table 3.2 ***Stereotyping the brown and green agendas of urban improvement***

	The 'brown' environmental health agenda	*The 'green' sustainability agenda*
Characteristic features of problems high on the agenda:		
Key impact	Human health	Ecosystem health
Timing	Immediate	Delayed
Scale	Local	Regional or global
Worst affected	Lower-income groups	Future generations
Characteristic attitudes to:		
Nature	Manipulate to serve need	Protect and work with
People	Work with	Educate
Environmental services	Provide more	Use less
Aspects emphasised in relation to:		
Water	Inadequate access and poor quality	Overuse; need to protect
Air	High human exposure to hazardous pollutants	Acid precipitation and greenhouse gas emissions
Solid waste	Inadequate provision for collection and removal	Excessive generation
Land	Inadequate access for low-income groups to housing	Loss of natural habitats and agricultural land to urban development
Human wastes	Inadequate provision for safely removing fecal material (and waste water) from living environment	Loss of nutrients and damage to water bodies from the release of sewage into waterways
Typical proponent:	*Urbanist*	*Environmentalist*

Source: McGranahan and Satterthwaite (2000).

As cities have grown and their activities have become more industrial, their demand for natural resources – in particular energy resources and water – has increased. However, this is not a direct or simple relationship. Hardoy *et al.* (2001) have pointed out that if a city's main means of energy production is fuelwood, urban growth can have major impacts on the adjacent forest resources. However, doomsday predictions of large-scale forest loss have generally not been proven; this suggests that as cities develop, an energy transition takes place away from fuelwood to cheaper and more appropriate sources such as electricity. Impacts on rural areas can still occur. For example, where rapid city growth means that fuel sources initially remain traditional, the urban residents continue to rely on rural supply sources such as wood and charcoal, perhaps pre-empting rural supplies. Where modern power sources have been developed they impact on rural areas; for example, dams force the relocation of rural populations and tap sources of water. However, Main (1995) questions the 'environmental demonology' of cities, presenting two main arguments. First, while cities are certainly the cause of negative environmental impacts on rural people and environments, they are also the cause of positive impacts. For example, through the economies of scale provided by concentrated populations, cities facilitate the switch to more abundant fuels such as electricity. Second, he argues that portraying cities in a negative light often overlooks the fact that had development taken place without urbanisation, then the environmental and other impacts would more likely have been far greater. The environmental impacts of the world's 6 billion people all trying to live off the land, growing crops, cutting fuelwood and so on, would be immense. He argues that changes in technology and consumption patterns are a more important determinant of environmental degradation than either growth or redistribution of the population. Indeed the movement of people to settlements may have alleviated the effects on rural environments. In the case of Africa, Main (1995) argues that there are three processes which mediate the relationship between urbanisation and rural environmental change. These are:

1. reducing rural population pressures on land and other environmental resources, as rural population growth rates decline and even in some regions become negative;
2. rising urban demand for rural resources, as urban population growth and industrialisation generate increasing consumption;
3. the spatial concentration of population and production.

Each of these processes has the potential for both positive and negative impacts. For example, declining rural population growth and in some cases declining populations imply reduced pressure on the land and therefore reduced rural environmental and land degradation. However, Main (1995) is careful to point out that the relationship between population and land degradation is not a direct one. For example, he cites the work of Boserup (1965) who argued that under circumstances of high population pressure new technologies or techniques are developed which can have positive environmental impacts. The work of Fairhead and Leach (1995; 1998) highlights the evidence that increasing rural population can lead to improved environmental management, in particular in the agroforestry systems where the orthodox research findings suggest that the reverse is true. Their findings suggest that the relationship between human activities and forest resources is far more complex than had been understood previously. For example, the inhabitants of their study area, Kissidougou, Guinea, make use of a wide range of forest resources. By adopting a range of forest management techniques, involving fire, soil, animals and vegetation, they are able to increase the biological diversity and the productivity of the forests, improving the value of their environment. Contrary to the orthodoxy, the study found evidence that there is a positive relationship between richness of the forest resources and population growth. In the title of their book, *More People, Less Erosion,* Tiffen *et al.* (1994; see also Mortimore and Tiffen, 1995) make explicit the fact that increased population leads to more careful and intensive environmental management. They provide evidence that the opportunity offered by the urban market was a crucial asset for the advancement of rural producers.

Main (1995) points out that the mobility of people in marginal rural environments has long been a strategy for sustaining livelihoods (see Chapter 4 below). This is an example of where the link with the urban environment is crucial to both rural livelihoods and rural environments. For example, he refers to the seasonal circulatory migration known in Hausa as *cin rani*; literally: 'eating away the dry season'. *Cin rani* involves spending months away from the rural home and earning money during the dry season when agricultural labour demand is low, thereby saving rural food stocks. However, from the mid-1970s the duration of the stay has tended to lengthen and the focus of migration has increasingly been the larger urban areas. During the 1972–74 drought this provided an obvious strategy by which rural dwellers could avoid the worst effects of this environmental hazard. Main (1995) suggests that the fact that the towns and cities were able to absorb large influxes of rural migrants during this period of intense environmental pressure is evidence of the effectiveness of the existing urban system and also of the effectiveness of the rural–urban linkages. If this migration had not taken place the environmental degradation is likely to have been worse and more people would have starved.

Hardoy *et al.* (2001) suggest a series of environmental advantages or opportunities available in cities:

- economies of scale and proximity to infrastructure and services;
- reducing risks from natural disasters;

- water reuse or recycling;
- access to land;
- reduced need for heating;
- reduced motor vehicle use;
- pollution control and management;
- funding environmental management;
- governance;
- potential for reducing greenhouse gas emissions.

In most cases, Hardoy *et al.* (2001) argue, these are *potential* environmental advantages offered by urban areas. For example, concentration of the population provides opportunities for what they call 'environmental economies of scale'. In most cases this relates to the declining relative costs of providing infrastructure or services as population density increases. However, the reality is more complex and challenging. The next sections review a number of key environmental issues that involve relations between rural and urban areas.

The issue of water

Particular problems for water-stressed countries arise where there are conflicts of interest over access to water. Most demand for water is for domestic urban and urban industrial uses, which compete with agriculture, while most sources of water are located in rural areas. The rapid growth of cities in developing countries is forcing the urban authorities to seek water sources at increasing distances. One of the implications of this is that the urban poor have to pay increasing prices to access water; in some cases the costs of obtaining water for the urban poor are higher than those for the affluent urban residents (see Figure 3.3).

As Burke and Beltran (2000) indicate, most cities have strong centralised institutions which are responsible for sourcing water, while rural users of water are far more diffuse and decentralised. An imbalance in power can result in which rural users lose out. However, as they point out, in most cases this is not necessary as the needs of the different users may not be incompatible. For example, most industrial and residential users require low volumes of high-quality water, while most agricultural users require high volumes of water and are generally indifferent to the quality. The authors go on to identify a series of initiatives that they argue are required to

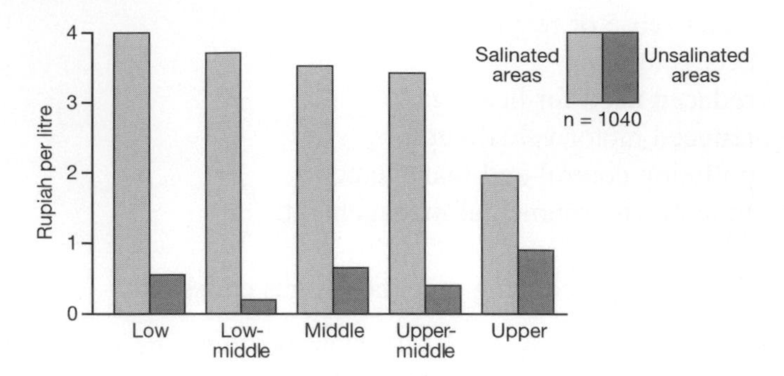

Figure 3.3 ***Average price paid for drinking water in Jakarta, by geographical features and wealth groups.***

Source: M. Kjellén and G. McGranahan (1997) *Urban Water – Towards Health and Sustainability*, Stockholm Environment Institute, Stockholm.

reconcile the differences between the interests of the rural and urban users:

- Economic appraisal of the opportunity costs of the water resource base
- Continuous examination of the role of economic instruments in water resource allocation for agriculture
- Assessment of source vulnerabilities in rural/urban settings where resource degradation has become apparent
- Clarification of the precise risk to public health posed by agricultural production and the use of effluent water in horticultural productions [see Box 3.3].
- The facilitation of structured negotiation between disparate user groups in both urban and rural areas
- The development of participatory planning procedures across relevant basins or aquifers, administrations and economic sectors
- The development of working systems of basin and aquifer governance where existing or customary arrangements have failed
- Evaluation of the role of crop management, irrigation technology and resource management approaches to manage the demand for raw water in agriculture (including advances in in-field application and drainage, the use of waste-water for irrigation, conjunctive use and aquifer-storage-recovery).

(Burke and Beltran, 2000)

One aspect of the problem of access to water is competing uses of water reservoirs. For example, Unkal Tank is a reservoir located on the main highway between the converging cities of Hubli and

Box 3.3

Sewage and farming in Hubli-Dharwad

In many peri-urban areas in the developing world it is common to find intensive cultivation of horticultural crops. This is the case because the adjacent urban area provides the market for the crops, the labour and the capital to invest in intensive cultivation methods. However, due to the high levels of demand for water in the city and the problem of pollution of the water courses, vegetables may be cultivated using contaminated water for irrigation. This problem can be compounded if the city is located in an arid or semi-arid zone where the demand for vegetables in the dry season can provide strong incentives to produce then, but low rainfall means that the dilution of water-borne pollution is far less.

In Hubli-Dharwad, a city-region of approximately 1.4 million people (Brook and Dávila, 2000) in semi-arid southern India, sewage is an extremely valuable resource for the peri-urban farmers. It provides a year-round, cheap source of irrigation, but is particularly important during the dry season, and the high nutrient loading can also reduce the need

Figure 3.4 ***Farmer near Hubli-Dharwad, India, adding endosulfan (a highly toxic contact pesticide) to his knapsack sprayer without protective clothing.***

Source: Andrew Mark Bradford, Royal Holloway.

for fertiliser inputs. Farmers report 30–50 per cent increases in productivity because of using sewage for irrigation. However, this is not without risk. For example, there is evidence that the sewage leads to a high level of weeds and pests, and soil and groundwater contamination, as well as to the danger of introducing sewage-borne pathogens and chemical toxins into the human food chain when either the farmer's family consumes the food or it is sold into the urban market and consumed there. Pests require treatment with organo-phosphates, but because farmers are often not trained to use them, it is common for them to apply these without any form of protective clothing (see Figure 3.4). The problem of contamination is rather starkly illustrated by the area near Hubli hospital: some of the sewage used originates from the hospital drain pipe and this has resulted in the contamination of land with biological waste. Bradford *et al.* (2003) reported interviewing farmers who complained of having trodden on disposable syringes and needles despite having fitted their pump inlets with filters (see Figure 3.5). The use of powerful chemicals has other implications. Bradford *et al.* highlight the fact that many of those employed by the peri-urban farmers in Hubli-Dharwad are women from the nearby urban areas. In their work the women may be exposed to the sewage and

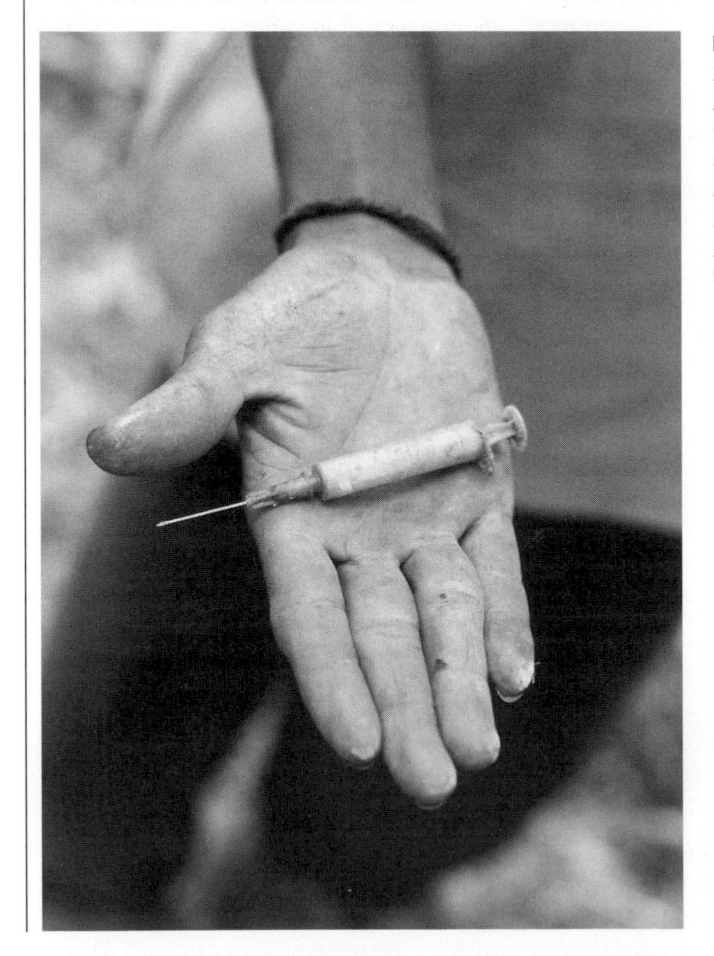

Figure 3.5
Farmers near Hubli hospital complain of treading on used disposable needles and syringes buried in the soil.

Source: Andrew Mark Bradford, Royal Holloway.

the chemical inputs, and this may pose a threat to their health. Also, many of these women will return home after a day's work to carry out their household chores such as preparing food, thus running the risk of passing pathogens on if basic standards of hygiene are not maintained.

These examples emphasise the importance of farmers developing strategies such as filtering their pump inlets. It also highlights the need for agricultural extension in urban and peri-urban areas, something that is often overlooked by ministries of agriculture that see their remit as largely rural-focused. For example, Bradford *et al.* (2003) reported that in their efforts to combat the pests associated with sewage-irrigated crop production, farmers are mixing pesticides, with the encouragement of pesticide dealers, producing potentially hazardous combinations. However, the pesticide dealers are the only source of agricultural extension information, so the farmers are given no other advice on such issues. In some cases farmers have been able to minimise the hazards to themselves and to the human food chain by engaging in strategies such as agro-forestry or fodder crop systems. These reduce the hazards by reducing the farmers' direct contact with and exposure to sewage because these cropping systmes require less irrigation. Bradford *et al.* suggest other alternatives that could be developed, including integrated pest management systems that avoid the use of these chemicals altogether and have the added benefit of reducing the cost of inputs.

Dharwad, in Karnataka State, India. The tank drains a watershed of 45 square kilometres much of which was developed for urban residential uses in the last decade of the twentieth century. It has a capacity of just over 2.25 million cubic metres. The tank was a key abstraction point for urban water supply, but in 1996 the quality was found to have declined to the point where it was not even suitable for treatment. This has meant that the tank has now been opened up for a range of leisure uses.

This change in use of the reservoir has meant that the range of stakeholders in the management of this water resource now includes (see Nunan *et al.*, no date):

- *dhobi wallahs* (laundry men and women), as well as local residents doing their own washing;
- Hotel Naveen, located on the shore of Unkal Tank;
- local fisherfolk, as well as a commercial fisherman;
- agriculturalists using water from Unkal Tank to irrigate their crops;
- the Hubli-Dharwad Municipal Corporation;
- the Hubli-Dharwad Urban Development Authority;
- the Karnataka Urban Water Supply and Drainage Board;

- the Karnataka State Town Planning Department;
- the Karnataka State Land Use Board;
- the Karnataka State Pollution Control Board.

The reservoir is thus now used by a range of stakeholders from across the urban–rural continuum. One approach to the management of these conflicting interests is to assign an economic value to the range of uses. Nunan *et al.* (no date) were able to assess the use value of Unkal Tank by calculating the cost of travel to the tank, the time spent travelling to and at the tank, and earnings derived from using the tank. The resulting values are summarised in Table 3.3.

Although this is crude, it provides a powerful analysis of the value of the tank to the various stakeholders and an indication of its (direct and indirect) economic value. The non-leisure value alone is estimated at 766,073 rupees (combining the categories in Table 3.3 but removing the religious uses from the washing category). This also indicates the potential economic loss if the levels of pollution continue to increase to the extent that some of the current uses become untenable. Nunan *et al.* (2000) were also able to attempt a valuation of the loss of Unkal Tank as a water supply by analysing the increased costs of sourcing water from elsewhere. Unkal Tank

Table 3.3 ***Unkal tank use values***

Use	*Value (Rs)*	*Breakdown (%)*	*Comments*
Recreation	12,303,639.00	90.2	Based on travel and hourly wage rates of users
Boating	558,450.00	4.1	Based on boating costs, travel time to tank and time spent boating
Commercial fishing	124,330.20	0.9	Based on fishery contract and associated incomes and costs
Individual fishing	138,687.16	1.0	Based on market value of catch and time spent fishing
Washing	423,813.13	3.1	Including washing clothes (commercial charges and time spent for local households), watering buffaloes and bullocks, bathing and religious ablutions
Washing vehicles	84,389.07	0.6	Including drivers of commercial vehicles and commercial vehicle washers
Total	13,633,308.56	100.0	

Source: adapted from Nunan *et al.* (no date).

was capable of providing 2.25 million cubic metres, while the current cost of supplying Hubli-Dharwad was estimated at 2.1 million rupees. Thus the cost of the loss of Unkal Tank is estimated to be 4.8 million rupees. In the meantime, water for the rapidly growing city is being sourced from other reservoirs further away at Malaprabha Dam and Neeragagar Tank, thus extending the city's ecological footprint. This approach provides an argument for improved management of this vital natural resource for rural, urban and peri-urban users. It also illustrates that a water body with a relatively high level of pollution still provides a high use value to its region.

Energy flows to cities

O'Meara (1999) argues that before the development of industrial energy production methods after the Industrial Revolution, available sources of energy limited city size to around 1 million residents, who were reliant on wind, water, wood, and human and animal power. With the advent of coal and steam-powered engines, large urban manufacturing plants became a possibility. With the emergence of the internal combustion engine, cities were able to expand beyond previous limits and to extend their influence beyond their boundaries. Later energy resources were brought in from more distant locations. However, there has been growing concern that as cities grow in size their consumption of energy resources has increased. This concern has focused in particular on urban areas in the developing world, where, despite the fact that their size may have grown beyond the 1 million threshold, residents are still often highly dependent on energy sources such as woodfuel. This has prompted much concern and research on what has become known as the 'woodfuel crisis'.

Mearns (1995; see also Leach and Mearns, 1988) argues that the 'woodfuel crisis' in East Africa represents the archetypal environmental problem of developing countries. His discussion of this indicates the importance of institutions in the management of the forest resource. This analysis is based on comparisons of per capita woodfuel consumption. Most developing countries where population is increasing experience a woodfuel supply shortfall. Since this has to be met from somewhere, Mearns reasons, it is assumed that it is met by cutting into tree stocks. This trend is then projected into the future, linked to population growth, pointing to deforestation, accelerated soil erosion and widespread environmental degradation. In addition, given that even in cities woodfuel is an important source of energy, it often has to be transported hundreds or even thousands of kilometres to urban-based consumers. For example, Hardoy *et al.* (2001) report that during 1981–82 Delhi consumed 612 tons of fuelwood per day that was brought in by train mostly from Madhya Pradesh (nearly 700 kilometres away), and also from Assam and Bihar. Delhi has a relatively low consumption of fuelwood because kerosene, coal and liquid petroleum gas (LPG) are also available. Far more fuelwood is consumed in Bangalore even though it is half the size of Delhi, because alternatives are not as widely available. Bangalore's supplies come mainly from private sources, such as farms and forests within 150 kilometres of the city, but it is estimated that up to 15 per cent is sourced from government forests between 300 and 700 kilometres away (Hardoy *et al.,* 2001). Figure 3.6 shows fuelwood cut from nearby forest, tied up and lying by the main road between Sokoto and Kano in Nigeria. Orthodox theory suggests that the growth of these two cities – particularly Kano – has prompted the search for woodfuel at increasingly distant locations, placing pressure on the environments of increasingly distant rural areas.

However, Mearns (1995) identifies four serious flaws in this analysis:

- Woodfuel consumption patterns defy generalisation. He illustrates this with data from a survey of 38 villages in Ethiopia where total annual energy consumption varies from 4 to 38 gigajoules per person, and is made up of a heterogeneous combination of woody biomass, charcoal, crop residue, oil products and dung.
- The data on which this kind of analysis is generally based are poor. These data can be based on vague estimates or proxy indicators which are not easily compared from one context to another.
- Mearns argues that it is unrealistic to assume that consumption will continue to rise in line with population, even while supplies dwindle. As scarcity increases, it is more likely that wood prices and labour costs will increase, prompting responses by both suppliers and consumers including planting trees, using fuel more economically, switching to more abundant fuels or encouraging natural regeneration of wood-producing vegetation.
- Finally, Mearns argues that it is clearance for agriculture that is the main cause of deforestation and not woodfuel consumption.

Figure 3.6 ***Fuelwood waiting to be purchased by passersby on a main route between Sokoto and Kano.***

These flaws, Mearns (1995) argues, prevent an appropriate response from governments and aid donors. He argues that it is an understanding of woodfuel within the context of people's livelihood choices that is the key to alleviating rural and urban energy problems. Indeed, he goes on to argue that institutional dimensions are key to understanding people's natural resource management practices. For example, in western Kenya, Mearns points out that despite abundant woody vegetation, surveys have indicated a household woodfuel deficit of approximately 633 kg per year per homestead. Woodfuel collection in this region is a female responsibility, but tree planting and control are mainly the responsibility of men. In this area many of the trees are managed in woodlots for the commercial production of poles for construction, again an activity largely controlled by the men. In addition, Mearns reports locally held beliefs that a woman who plants a tree will become barren or her husband will die or she is making a direct challenge to his household supremacy. Such cultural influences on the management of natural resources are subject to variation between societies, according to household circumstances and across time, but are key to an understanding of the complex social and cultural relations that influence the management of natural resources such as wood.

Another key issue raised in the literature on woodfuel is the connection between woodfuel scarcity and price variation. Conventionally, one might expect prices to rise as woodfuel stocks are depleted. Woodfuel or charcoal prices have therefore been monitored as an indicator of the woodfuel crisis. However, long-term analysis of charcoal prices in Khartoum, Sudan's capital city, indicate a pattern of large fluctuations within years and from one year to the next, but with no obvious long-term trend. On closer analysis, Mearns argues that variations in labour availability turn out to be a more important explanatory factor. Approximately 80 per cent of Khartoum's charcoal comes from Blue Nile and Kassala provinces, where its production is an alternative to agricultural labour, so when agricultural labour is in high demand labour for charcoal production becomes scarce. In addition, when petrol prices are high, so are transport costs, which are passed on to the Khartoum-based consumers through an increase in the price of charcoal. Prices are then further raised by the increased demand for charcoal from urban residents unable to afford petroleum-related products such as kerosene. Price in this case is therefore a function of a more complex set of circumstances than simply scarcity or abundance of supply.

Mearns (1995) argues that appropriate institutions for managing the environment and protecting those vulnerable to environmental impacts are absent. This leads him to argue that the 'woodfuel crisis' is less a problem of the physical scarcity of woodfuel and more a problem of 'institutional scarcity', a conclusion that is shared by Iaquinta and Drescher (2001). In seeking policies for solving this problem, improved institutional structures are required, not least for improving the structure of urban woodfuel markets and improving an understanding of the dynamics in operation between the various energy sources.

One way in which it may be possible for existing institutions to manage the land to maximum effect is proposed by Nunan *et al.* (no date) and shown in Figure 3.7. This figure illustrates how three main use values can be applied to the land on the edge of cities but strongly influenced by urban economies. The three key values are:

- *direct use value* – the use value normally applied, that is the costs or benefits earned from direct use of the natural environment;
- *indirect use value* – the costs and benefits derived from support and protection provided to activities and assets by natural resources;
- *non-use value* – the benefits and costs involved in not using natural resources, for example the protection of wildlife or conservation areas.

Agriculture is perhaps the simplest illustration of direct use value to which natural resources can be put in the case of agricultural land – whether peri-urban, which was Nunan *et al.*'s focus, or rural. The use value is therefore calculated from the benefits derived directly from consuming the produce or from the money raised by selling the produce. Indirect use values can be derived from the benefits of forestry for protection of watersheds or soil resources. Agricultural land in peri-urban areas can also provide the indirect use value of protecting urban land from risks of flooding (see for example Lynch *et al.*, 2001). One subcategory of indirect use value is what Nunan *et al.* (no date) describe as *option value*. This refers to the potential direct or indirect use value that may occur in the future.

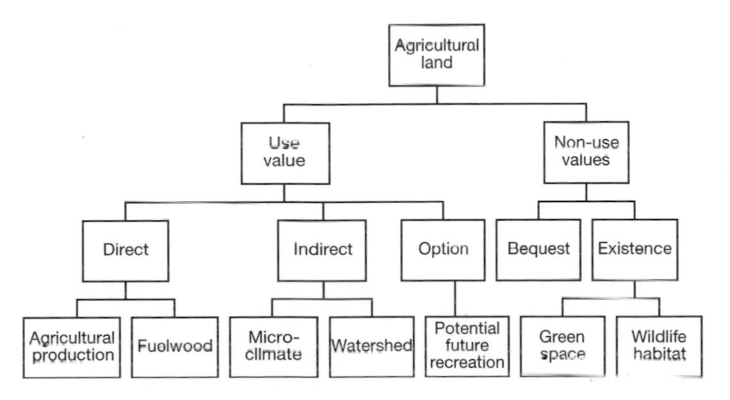

Figure 3.7 ***Total economic value for agricultural land.***

Source: F. Nunan, K. Bird and J. Bishop with A. Edmundson and S. R. Nidagundi (no date) *Valuing Peri-urban Natural Resources: a Guide for Natural Resources Managers.* School of Public Policy, University of Birmingham.

For example, some have suggested that urban and peri-urban agriculture could be used as a planning tool if protected from competing land uses. Then, as the surrounding area becomes built up, the land could be allocated for important urban services such as schools, community centres and roads (Sawio, 1994). This is similar to the system of greenbelts in the UK where planning protection is needed for such land as urban development takes place and puts it under increasing developmental pressure. Finally non-use value is derived from protecting natural resources for their own sake, for example as habitat for wildlife. The challenge here is that these values may conflict or overlap between different groups of users, making distinguishing these categories extremely difficult.

Peri-urban environments

One of the most profound implications of the development of urbanisation is the effect it has on surrounding rural areas. In the literature this surrounding area which is strongly influenced by urban activities is often referred to as the 'peri-urban interface' (Brook and Dávila, 2000) or 'peri-urban environment' (Iaquinta and Drescher, 2001). Others have termed it the 'Extended Metropolitan Area' (Pacione, 2001) or, in the case of Asia, the *desakota* (McGee, 1991). This is discussed in more detail in Chapter 1 above. Allen (2001) argues that the peri-urban environment has a unique problem of institutional fragmentation as a result of overlapping and changing local government institutions which are either still rural or have developed into urban (illustrated in columns two and three of Table 3.4) but are inappropriate for peri-urban. Iaquinta and Drescher discuss the rural, peri-urban and urban forms in terms of a linked system, which is an 'uneven or *lumpy*, multidimensional continuum' (2001: 3; emphasis in original): they are suggesting that the peri-urban continuum is far from a smooth, linear transition from urban to rural and that it is subject to a range of influences and processes because of its location at the interface between urban and rural areas. They suggest that it is possible to identify a peri-urban typology (illustrated in Table 3.4) on a continuum from edge-of-city to urban-influenced rural locations. They argue that such an understanding is important for environmental management of these areas.

Iaquinta and Drescher's (2001) typology comprises five ideal types that are embedded in the broader rural–urban dynamic. In this sense

Table 3.4 ***Summary of peri-urban typology and accompanying institutional contexts***

Description	*Geographical dimension*	*Migration dimension*	*Institutional context*	*Influences*
Village peri-urban	Distant from the city both geographically and in travel time	Derives from sojourning, circulation and migration	Embodies a *network-induced institutional context* wherein change is effected through diffusion or induction while institutions remain in orientation and stable.	• Inflow of out-migrant remittances • Out-migrant infusion of 'urban' ideas and modes of behaviour • Out-migrant infusion of non-income resources • Out-migrant participation – particularly strategic – in community decision making
Diffuse peri-urban	Urban fringe	Derives from multiple-source point in-migration	Embodies an amalgamated institutional context	High-level ethnic heterogeneity because of multiple-origin in-migration, with greater density of various beliefs about customary institutions leading to increased likelihood of conflict and the possibility of the development of new institutions for resolving conflict.
Chain peri-urban	Urban fringe	Derives from chain migration	Embodies a *reconstituted institutional context* where links to the donor area remain strong and traditions and institutions are transplanted with some modification from the donor area and take on a slightly defensive character	High level of homogeneity, with integration of urban institutions with traditional and customary institutions
In-place peri-urban	Close to the city; urban fringe	Derives from in-place urbanisation, natural increase and some migration	Embodies a *traditional institutional context* with long-term stable institutions evidencing strong defensive insulation	Least likely to be opportunistic, since the residents have chosen to remain, most likely to benefit from traditional arrangements; most likely to have newcomer–oldcomer polarisation, resulting in conservative institutions
Absorbed peri-urban	Within the city, having been absorbed	Derives from succession/ displacement and traditionalism	Embodies a *residual institutional context* wherein the roots of social arrangements lie in the traditions of a previously resident culture group and are now maintained through ritualism	Deriving from either in-place or chain areas, resulting in relatively conservative institutions despite the settler culture being replaced through residential succession or diffusion due to migration

Source: adapted from Iaquinta and Drescher (2001).

they see the peri-urban interface as being crucial to the mediation of the linkages between rural and urban areas. They argue that in each case migration is a key to understanding the links between the elements (see Chapter 4 below). This is in the context of empirical research they have carried out on the role of peri-urban areas for the supply of food (see Chapter 2 above) and environmental management and planning (Iaquinta and Drescher, 2001). In this sense, they suggest that, in addition to an environmental footprint, a social footprint can be identified. This is crucial to understanding natural resource management and planning, and therefore the mediation of the ecological footprint.

They suggest that one of the key issues to consider is that these typologies are linked in either time or space, through migration – bringing influences across space – or urbanisation – making changes over time. In this context they define urbanisation broadly as a process comprising two phases.

- The first phase involves the movement of people from rural to urban places where they engage principally in non-rural urban activities.
- The second phase relates to a change in lifestyle from rural to urban, involving changes in attitude, values and behaviours (see Chapter 5 below).

These two phases are mutually supportive, with the former manifested by changes in population density and economic function, while the latter is manifested more by behaviours and social and cultural factors. In this context urbanisation is not only a question of the development of the city out into the rural area, changing the rural into the urban. Iaquinta and Drescher (2001) argue that it is important to see geographical location as a dimension that *aids* the definition of different types of peri-urban space, rather than determining a definition of it. This is illustrated in the multi-stranded urban–rural continuum in Figure 3.8. It is important to see the linkages that take place in the peri-urban as dynamic, interactive and transformative. This means that they are places which are influenced from outside and transformed by influences of transition. For example, urban agriculture is increasing in many cities (see Chapter 2 above), while non-farm employment is increasing in many places in the countryside.

An understanding of such linkages is required in order to manage the peri-urban environment. Human networks are often referred to as

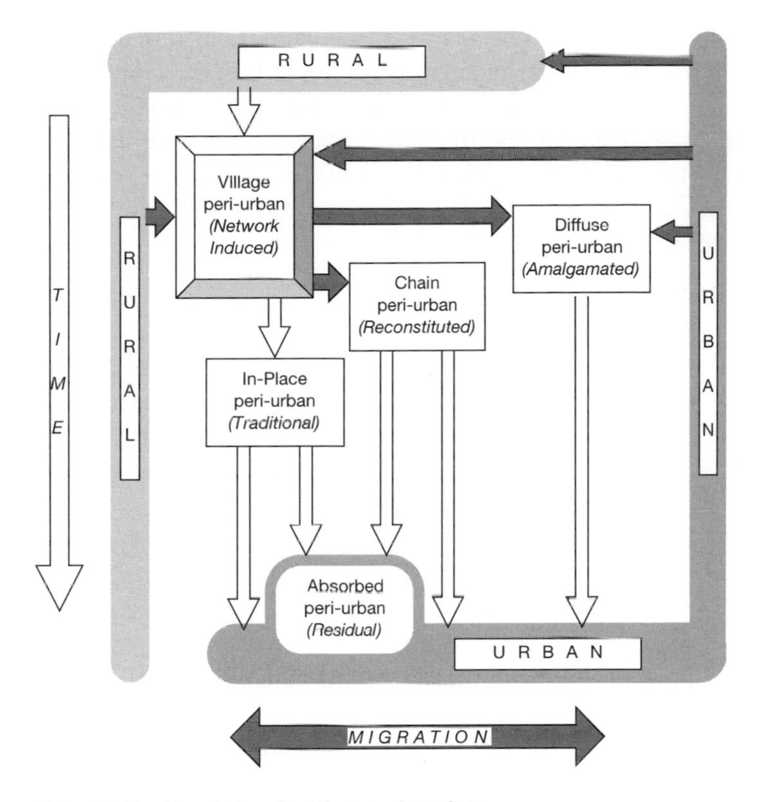

Figure 3.8 ***Typology of peri-urban interface.***

Source: D. Iaquinta and A. Drescher (2001) More than the spatial fringe: an application of the peri-urban typology to planning and management of natural resources. Paper presented to the conference: Rural–Urban Encounters: Managing the Environment of the Peri-Urban Interface. Development Planning Unit; University College London.

'social capital' because they can provide a form of livelihood asset that is made up of social relations – an equivalent to economic capital which is built up through economic relations. An understanding of the social capital networks of the peri-urban areas will provide insights into the key influences on how the environment is managed. For example, institutions are central to the issue of how land tenure is mediated. This is where the concept of the social footprint becomes important to an understanding of the ecological footprint. Iaquinta and Drescher (2001) argue that institutions and policies must be balanced between those that encourage micro-household exchange and those that encourage exchange between generations within households. This is important in order to ensure security of tenure, to maximise farmers' choices and to discourage moves from intensive to extensive modes of agricultural production, with the related environmental implications.

This emphasis on the issue of the institutional framework for environmental management is reinforced by the University of London's Development Planning Unit (no date) which prepared three booklets on environmental management and planning in the peri-urban interface. They argue that there are three characteristics that make planning and management of the peri-urban distinctive. These are:

- **Changing locations.** As the city or town expands its influence rural areas constantly become part of the peri-urban area as former peri-urban areas become part of the urban area. The peri-urban is characterised by a mix of both urban and rural, incorporating land uses such as landfills, mining, industrial developments or airports. However, the changing locations make responsibility difficult to define.
- **Changing populations.** People are affected by peri-urban interface change as migrants arriving from both urban and rural areas mix with those already *in situ*. This means that networks of actors and institutions are dynamic, making institutions impermanent and the consistent management of natural resources concomitantly problematic.
- **Weak and overlapping institutional structures.** Because of the range of competing interests in the peri-urban interface, the institutional structure is very fluid and usually unable to provide a balance between them. Thus the poor and the environment are often adversely affected, with little power or voice in such a context.

It has become clear that the location and space of the so-called interface between the urban and the rural are less important than the linkages that are mapped across it. Location in social, economic and political networks is more important than physical proximity to urban or rural areas. In particular, Iaquinta and Drescher (2001) highlight migration and the changes in ideas, lifestyles and livelihoods available to those who live in this interface as being crucial.

Conclusion

This chapter has discussed a number of environmental aspects to the linkages between cities and the countryside in developing countries. It has reviewed the approaches to researching this topic which have emphasised that the environmental burden of urbanisation does not only fall on those living in urban areas. Indeed, there is an argument that some cities, as they become more affluent, are able to distance their environmental burden in both space and time.

Most urban water supply sources are in rural areas. We found that as cities grow in size and affluence, the distance from which water and other resources are drawn increases. However, in a discussion of the issue of energy we found that crude links between population growth (and levels of urbanisation) and forest degradation were based on faulty analysis of the problem. There are clearly fuelwood problems in some areas, but simple population growth is not the cause.

Finally, two approaches to understanding environmental linkages are discussed. The first focuses on the institutional structures that mediate the urban–rural interface. While Iaquinta and Drescher (2001) are mainly concerned with the peri-urban interface they define this in broad terms, identifying five typologies and their varying institutional structures. The second approach emphasises attempts to place an economic value on natural assets in order to build in a costing for all externalities.

Discussion questions

- Review the main ways in which the cities of the developing world impact on rural environments.
- Critically assess the extent of the urban woodfuel crisis in sub-Saharan Africa.
- Debate the case for considering peri-urban areas as unique and in need of specialist environmental planning and management strategies of their own.

Suggested reading

Lynch, K. (2004) Managing urbanisation. In F. Harris (ed.) *Global Environmental Issues*. John Wiley, Chichester. 195–228.

McGranahan, G. and Satterthwaite, D. (2002) The environmental dimensions of sustainable development for cities. *Geography*. 87 (3), 213–226.

Main, H. (1995) The effects of urbanisation in rural environments in Africa. In T. Binns (ed.) *People and Environment in Africa*. John Wiley, Chichester. 47–57.

Rees, W. (1992) Ecological footprints and carrying capacity: what urban economics leaves out. *Environment and Urbanization* 4 (2), 121–130.

Wackernagel, M. (1998) The ecological footprint of Santiago de Chile. *Local Environment* 3 (1), 7–25.

4 People

Summary

- We are living at a time when some cities are growing at rates faster than have ever previously been seen.
- Much of the research on rural–urban interaction focuses on the movement of people and is dominated by a consideration of rural-to-urban migration.
- More recent research has identified a variety of movements between rural and urban areas, including step-wise migration (village – town – city), circulatory migration (village – city – village), cyclical migration (associated with seasonal variation in labour demand), multi-locational households (where households have members in town and country) and chain migration (where migrants follow their predecessors, and are assisted by them in establishing an urban base).

Introduction

Much research on urban areas and the relationships between urban and rural areas in the developing world is focused on the rapid population growth of cities. Past research on both urban growth and rural poverty has focused on the role of the movement of people into cities. This chapter examines the flows of people between the urban and rural areas. It discusses the motivation behind people's movements and their place in strategies for maintaining livelihoods.

Cities in the developing world

Ever since we have had records about the development of cities, a major concern has been how to cope with urban growth. However, as Potter and Lloyd-Evans (1998) point out, we are living during a period in which the number of people living in cities around the world is growing faster than at any previous time. Most urban population growth is taking place in the developing world. The 1994 International Conference on Population and Development (ICPD) reported that about half of the world's governments considered their population distribution to be unsatisfactory and wanted to change it in some way (Johnson, 1995). In most of these cases the focus of concern was the speed of growth of the cities in the developing world. This chapter focuses on the flows of people across the urban–rural interface.

The population growth described above has led to a twofold change in population distribution. First, there is an absolute growth in the number of people living in cities and towns; this is urban growth. The second relates to the relative change in the proportion of people living in towns and cities as opposed to rural areas. This is referred to as urbanisation, the shift being from country to town (counter-urbanisation is the shift in the opposite direction). The terms themselves suggest an urban-biased perspective. The concept of urbanisation is used to describe the relative shift of a country's or region's population towards greater urban concentration. Urbanisation is also used to refer to other related processes of the change in the economy, culture and society, where urban processes, systems or influences begin to dominate. These latter dimensions of urbanisation are discussed in other chapters in this book.

The challenge of managing rapidly growing urban environments was identified as early as 1980 by the Brandt Report, more formally known as the report of the Independent Commission on International Development Issues (1980) *North–South: a Programme for Survival.* This concern was emphasised in *Our Common Future*, produced by the World Commission on Environment and Development (1987), and also known as the Brundtland Report after the chair of the Commission, Gro Haarlem Brundtland. The irony of these two international reports focusing on the concern that the developing world would become more urbanised over the next 20 years is that until relatively recently it was widely accepted that urbanisation was associated with development, modernisation and economic growth.

In order to understand how this reversal has come about, it is necessary to briefly revisit the early formation of cities, and understand their international as well as national role.

In sub-Saharan Africa, before the period of colonisation, people showed evidence of a high degree of mobility. Much of the history of pre-colonial civilisations involved the movement of people across the continent. During the period of colonisation, the imperial powers established settlements that were able to service the industries that they were seeking to develop in their colonies. As shown in Figure 4.1, many of the countries of Africa and Latin America now have capital cities that are located on the coast. This is particularly evident along the coast of West Africa. During the pre-colonial period the important cities in this region were inland, including Djenne (now in Mali), founded in the ninth century, Gao (Mali) and Kano (Nigeria), dating from the tenth century, and Zaria (Nigeria) and Timbuctu (Mali), from the eleventh century (Potter *et al.,* 2004). The colonial period established a number of the major cities as colonial port-capitals, such as Lagos (Nigeria), Dakar (Senegal) and Accra (Ghana). In Latin America the major present-day cities were founded by the colonisers, for example Rio de Janeiro and São Paulo (Brazil), Lima (Peru), Buenos Aires (Argentina), Santiago (Chile) and Caracas (Venezuela). In Asia, cities such as Kalkot and Mumbai (India) and Colombo (Sri Lanka) were European coastal port-cities and have since become major metropolitan centres. It was only in the Middle East and China that the pre-existing urban hierarchies were strong enough to resist the imprint of the European colonising era.

Many of the colonial cities were entirely new settlements or established on the site of pre-existing settlements. Most were port-cities, or located close to ports. These settlements served as ports, administrative and military centres, effectively gateways and capitals to the colonies. In order to simplify administration and governance the colonialists concentrated their seats of power in one or a small number of centres. The most obvious location of these centres was on the coast, where the ships of the metropolitan core arrived to collect the natural resources being produced. A quick glance at a map of Africa or Latin America highlights this, with the majority of the contemporary capital cities being former colonial capitals located at ports. For many colonies, this represented the beginning of a form of urban dominance. The implications of this change in the spatial economy were that the economic lives of the people of Africa, Asia and Latin America began to revolve around the European-controlled mines, plantations and port-cities on the coast (see Figure 4.1 for maps of Latin America and Africa).

During the early period of independence, many former colonies began to develop their capital cities, attracting companies and diplomatic missions. They established ministries and government offices, all of which reinforced the pre-eminence of the colonial capital and often resulted in a condition of urban primacy – a situation where the main city dominates a society. This is illustrated by the example of Tehran in Iran (Mandanipour, 1998). In the mid-1970s Tehran accommodated 13.3 per cent of Iran's population and 28.6 per cent of the country's urban population, but it:

- accounted for 72 per cent of migration between provinces and 44 per cent of migration between urban areas;
- produced half of the GNP (excluding oil) revenue;
- accounted for 40 per cent of national investment;
- absorbed 60 per cent of industrial investment;
- housed 40 per cent of the large industrial concerns;
- accounted for 40 per cent of retail employment;
- had 56.8 per cent of hospital beds;
- accounted for 57 per cent of physicians;
- accounted for 64 per cent of newspaper distribution;
- had 68 per cent of vehicle registrations;
- had a population whose income was 70 per cent higher than in small towns and 45 per cent higher than in other large cities.

These income and other disparities not only between urban and rural areas, but also between high- and low-income urban residents, may have led to the tensions in the late 1970s that resulted in the Islamic Revolution in 1979.

In other countries, colonial powers placed very strict controls on urban migration by indigenous people in order to prevent the cities developing too rapidly and placing a burden on the colonial economy. In South Africa and many other African colonies, this resulted in Pass Laws, where the black African population were required to carry a pass giving them permission to be in certain locations. This usually excluded them from living in cities unless they were formally employed. In the case of South Africa, black South Africans and Africans from other colonies were encouraged to seek paid employment through various measures such as a hut tax, poll taxes and the banning of traditional subsistence agricultural

Figure 4.1a Map of Africa.

systems. Blacks working in a South African city had to have a pass to be legally allowed there. The only way of obtaining a pass was from an employer. Colonialism therefore established a system where migrant labourers were temporary urban dwellers, in town solely for employment, while keeping their permanent home in the village (see Box 4.1). Housing for the migrant workers was restricted to certain areas of the town or city which were often allocated to men only (Lester, 1998). The employee's family, therefore, survived on subsistence agriculture and the remittances he sent home (in most cases the only jobs available were for men). The benefit of this to the South African industries and government was that they had a smaller urban population and employees were always keen to return home and therefore unable to take up a permanent contract. The effect on the black South Africans was that their families became fragmented

Figure 4.1b ***Map of South America.***

as the men were away on three- and six-month contracts, while the women were left to maintain the household and keep the farm going. Such a mobile and temporary labour force ensured that the cost of labour was kept low and it was very difficult for workers to organise into any forms of resistance. Resistance to colonialism was beginning to emerge in other colonies and this was largely organised through labour groups, student groups and unions and professional associations. In many cases this focused on urban elites – another reason for the colonial powers to restrict the mobility of the indigenous population and to prevent the establishment of a large urban population.

As the colonial period in Africa ended in the 1950s, these restrictions were lifted, resulting in unprecedented rates of urban growth which have been sustained up to the present day. There is evidence of slowing urban growth during the 1980s, which Potts (1995) attributes to the economic hardship of this period. This apparently clear story has resulted in an orthodoxy that people in Third World countries tend to move to cities (in contrast to the counter-urbanisation of developed economies). However, there is growing evidence that suggests that the movements are more complex. One major difficulty for those studying the movement of populations in the developing world is the lack of accurate, reliable and timely data on which to develop interpretations of flows of people (Findlay and Findlay, 1991). For example, Briggs (1993), in a short paper, highlights some of the difficulties associated with the 1989 census in Tanzania, and the consequences of this for interpreting the total population of a large city like Dar es Salaam, let alone any more complex issues such as the flows of people into and out of the city. Tiffen (2003) also summarises some of the key challenges involved in making any generalisations from census data.

Contemporary migration

The explanation for rural–urban migration is complex and varied. In South-East Asia patterns of rural–urban migration are evident. The processes promoting this trend are varied. Drakakis-Smith (1996) reports that land reforms in many Asian countries have contributed to rural–urban migration in countries like Taiwan and South Korea, whereas in Thailand urban-based growth has not been linked to land reforms (Rigg, 1991). By contrast, in Bangladesh rural poverty has not led to rural–urban migration to the same extent as reported elsewhere.

Drakakis-Smith (1996) speculates that marked improvements in transport and communications have played a vital role in informing remote rural dwellers of the opportunities of urban lifestyles (see Chapter 5 below). However, if this were the case one would expect much migration to focus on the dominant, or primate, cities. There is

Box 4.1

Temporary migration as a survival strategy

A number of studies have recently tried to move away from an approach to understand migration that is based on the idea that populations are naturally settled and that when they move in the main the intention is to move permanently. De Haan (1999) suggests that migration studies have been pervaded by a perception that is based on the historical European demographic transition from peasant production to urban industrialisation. In developing countries, de Haan argues, the largest proportion of migrants move between rural areas. Many policy makers in the second half of the twentieth century endeavoured to restrict rural-to-urban migration, as it was feared that this had an adverse effect, mainly on the urban areas. China enacted a system of keeping the rural population in their country areas (see Box 4.2). However, de Haan points out that in China before the 1949 revolution the population was already highly mobile, but this did not lead to urbanisation.

The effects of rural-to-urban migration caused great anxiety in countries that, after achieving independence, lifted colonial laws restricting movements of the indigenous population. 'Views about migrants are often based on an assumption of sedentarism, that populations used to be immobile and have been uprooted by economic or environmental forces' (de Haan, 1999: 7). For example, migration in South Asia may in fact have decreased. In nineteenth-century Bengal large seasonal migrations took place during harvest times to the east, to the forests of the south and to the rice fields of the north-east. Large-scale migration from Bihar to Kalkot (Calcutta) still continues because of relative poverty, but it dates back at least as far as the middle of the nineteenth century.

In many regions of the world these migrations date back to colonial times, when migrant labour was cheap and easy to manage. In addition to the Indian examples, a highly mobile population has been a feature of Southern Africa for as long as mines have needed labour. Present-day South Africa still has a high number of migrants from neighbouring countries such as Malawi and Zimbabwe, as well as a highly mobile indigenous population. Rigg (1998b) reports a number of rural areas in South-East Asia where non-farm income is now an integral part of rural livelihood strategies. Many households have a family member working outside the village in non-farm employment. According to a village study in central Java, this is seen by many of the villagers as a matter of survival. In a survey of a range of village studies in south-east Asia, Rigg finds evidence that between 30 and 50 per cent of household income is derived from off-farm employment. The result is that in rural areas the relationship between farm income and poverty is less strong. One of the consequences of this is increasing differentiation between rich and poor households in some parts of rural Thailand. De Haan's (1999) arguments provide a possible explanation. Migration is selective and provides greater returns to those better able to take advantage of the opportunity. Second, there are costs and barriers associated with migration, which steer gains towards the better off. 'The chief question is not about migration itself, but what kinds of opportunities are available for what groups of people, and whether the type of migratory work allows the migrants and their families to improve their assets and "human capital"' (de Haan, 1999: 27).

evidence that much of the urban growth experienced in many of the world's poorest countries affects the smaller and intermediate-sized settlements. Pacione (2001) points out that while Latin America now has some of the largest cities in the world a number of medium-sized cities have emerged as powerful economic forces. He also argues that in Africa, where the urbanisation rates are highest, medium-sized cities are growing at least as fast as the largest cities such as Lagos and Cairo. While during the early post-colonial period much of the urban growth was caused by rural-to-urban migration, natural growth is now accounting for a greater proportion. Much more research is needed to provide a clear picture of the nature of population change, including features such as the relative importance of smaller and primate cities, rural-to-urban migration, natural population growth and migration in steps up the urban hierarchy in what is often described as step-wise migration.

The key visible indicator of urban primacy is the concentration of population in the dominating city. For example, it was reported at the International Conference on Population and Development (ICPD, 1994), that in 1992 there were 13 cities with a population of over 10 million. In itself, this is a cause for concern, but around half of these cities were in advanced economies. The factor that raised particular alarm about the growth in 10-million-population cities was that nearly all the new ones were predicted to be in less advanced economies, where they dominate the urban hierarchies and where there are fewer resources available to manage such huge population agglomerations. For example, Youssry and Aboul Atta (1997) point out that the population of Egypt grew more than fivefold in the twentieth century, while its capital, Cairo, grew more than sixteenfold. It grew from 600,000 in 1900 to 12.45 million by 1994; it accounted for 9 per cent of the population in 1940 and 21 per cent by 1994 (Youssry and Aboul Atta, 1997). However, Youssry and Aboul Atta (1997) also point out that population growth in some rural areas has been faster. One explanation for this is that the high cost of housing in Cairo has prompted people to build new houses in rural areas close to the city, resulting in an estimated 2 million daily commuters. Rigg (1991) reported that Bangkok had a population estimated between 100,000 and 400,000 in 1900. The city grew at rates of up to 8 per cent per annum (doubling every ten years), achieving its first million by 1950 and its second by 1970. By 1988 it was just under 6 million and by 2000 the urban agglomeration around and including Bangkok had a population of 8.76 million according to the United Nations Economic and Social Commission for Asia and the Pacific (UNESCAP, 2001). This means the city at its greatest extent accounts for 14 per cent of the population of Thailand.

Such headline figures are relatively easy to estimate and are persuasive. Urban primacy is therefore often measured on the indicator of population. For example, there are urban primacy indices which compare the population of the largest city against the second largest or the next four largest cities. However, while population data provide quantitative indicators of the dominance of a city, primacy also relates to other aspects of the relationship between the city and the rest of the country. Primacy indices are based on the relative differences between the population of the main city and of other cities. This gives no indication of the relationship between the city and the countryside. The focus on population omits measurement of any other aspect in which the main city dominates, such as those referred to above in the case of Tehran and concentrations of wealth, power and social and cultural outlets.

Some authors argue that orthodox models of migration assume that the flows are from rural areas to urban areas (Chant, 1998; Jamal and Weeks, 1988; Meagher, 1997; Potts, 1995). They provide evidence from various viewpoints that such models are inadequate, because there is evidence of urban–rural migration and changing household structures. Migration cannot be treated as a transition from 'rural' to 'urban', given that some households deliberately straddle this 'divide' as a livelihood strategy.

Much of the basis of research that examines population in relation to the rural–urban interface relates to the migration of people across that interface. In turn, much of this research focuses on rural-to-urban migration. To some extent this is understandable, since net migration flows tend to be in this direction. However, there are questions about some of the research that has been carried out. For example, Feldman (1999) argued that previous approaches to the study of rural–urban migration were flawed, for reasons including the following:

- They focused on seasonal or temporary migration, rather than on permanent migrants.
- They tended to focus on individuals rather than on households, thus concentrating on migration as an outcome resulting from rural or urban change, rather than as a process.

- There was a lack of research on the impacts of out-migration, such as the loss of agricultural labour capacity, the impact on rural infrastructural needs and the impacts of increased incomes on consumption or investment.

The result is an assumption that the remaining rural population is primarily dependent on remittances from urban migrants. This assumption has meant that non-farm employment and micro-industries in rural areas have received little research attention until relatively recently (see, for example, Bryceson, 1999, and Rigg, 1998a, 1998b). Only recently has it been possible to examine the highly varied demographic profiles of migrants and their complex decision-making processes (see for example Chant, 1998, on migration and gender, de Haan, 1999, on migration and livelihoods). In particular, research which focuses on the social networks of migrants has provided powerful analyses of migration processes which are linked to more complex understandings of informal-sector activities, slum and squatter communities, rural communities and the flows of people, capital and ideas through the social networks: 'What kinds of continuity and discontinuity shape information and resource flows between [migration] sites? What kind of kin and familial networks facilitate and hinder mobility?' (Feldman, 1999: 6; see also Box 4.2).

Patterns of rural–urban migration

The 2003 *State of the World Population* highlights the fact that rural–urban migration varies significantly in its advantages and disadvantages across the world. It gives as an example migrant textile workers in Nigeria and Bangladesh:

> The experience of rural-to-urban migrants varies considerably. In many developing countries, domestic work is one of the main sources of income for girls and young women in urban areas. In Bangladesh, textile work in cities has offered young women migrants unprecedented opportunities to earn money, save for dowries and postpone marriage; most of their experience has been very positive. In Nigeria, in contrast, young women apprenticing to be tailors are very vulnerable to sexual abuse because of their subordinate position at work and separation from their families.
>
> (UNFPA, 2003: 4)

Box 4.2

Migration and social networks

As an example of an in-depth study on rural–urban migration, Schneider (1999) provides some interesting findings on the link between livelihood strategies and migration in a comparison of data from Chiang Mai in Thailand and Baguio City in the Philippines. The two cities have quite different histories; both are second cities to primate capitals, Manila and Bangkok, both are similar in demographic size and both play a similar role in central place function in their region. However, Baguio City has experienced relatively high in-migration, while Chiang Mai received relatively low – possibly negative – net in-migration.

Schneider carried out a survey of residents and found that the reasons for migrating into the cities were predominantly economic – 76 per cent in Chiang Mai and 60 per cent in Baguio City. Schneider sought data on migration partners, hypothesising that the nature of the relationship with the partner would provide evidence of how the migratory move is embedded in social relations. In each case almost 60 per cent of the migrants moved with someone else, usually a family member, but of the 20 per cent of Chiang Mai respondents who moved to the city with a friend, nearly three-quarters were female. Women accounted for a relatively low percentage of the migrants who moved to the cities alone. However, Schneider finds the trend over time to be most interesting: in both cities the proportion of family/kin migration partners has declined over time. This Schneider interprets as an indication of social change paralleling urbanisation, which increases the importance of market forces in the regulation of livelihoods. This results in the development of new social networks increasingly intended to maximise livelihood opportunities, meaning that the decision to migrate temporarily or permanently is part of the livelihood strategy of households and families; consequently migrants leave families behind in search of improved income-earning opportunities.

This pattern whereby migrants make use of their social networks is reinforced by the data in Table 4.1 illustrating that Schneider found that large percentages of the migrants lived with relatives when they first arrived (62.5 per cent in Chiang Mai and 43.5 per cent in Baguio City), and most (55 per cent in both cities) found work within a few days. Schneider suggests that often relatives in the city can mediate on behalf of migrants for their first employment; thus the social networks become important for accommodation and employment. However, the evidence suggests that family and kinship ties as the main support relationship are in decline in favour of networks that may have been developed through previous migration experiences (see Box 4.3). Evidence from other studies seems to reinforce this.

Table 4.1 ***Category of first accommodation of migrants on arrival in Chiang Mai, Thailand, and Baguio City, the Philippines (%)***

	Baguio City	*Chiang Mai*
Alone	13.5	18.0
With friends	18.0	30.5
With kin	62.5	43.5
With others	5.5	8.0

Source: adapted from Schneider (1999).

Trager (1996) argues that research on rural–urban migration has gone through three phases. First, anthropologists were interested in the way rural in-migrants adapted to city life. Trager identifies even in this early work in the late 1960s, what she suggests is discomfort with the apparent separation of the rural and urban social fields, suggesting that many social networks cut across these boundaries. This nuance was sidelined by work in the late 1970s and 1980s which characterised urban and rural settings as being not only separate but competing (see in particular Lipton, 1977). Some, however, maintained that the linkages between the two realms were important to the understanding, and possibly even the development, of both. For example, McGee (1971), in his analysis of urbanisation in South-East Asia, argues that there is a rural–urban continuum. O'Connor (1983) argued that in Africa, more than any other region in the world, people live in a system of economic relationships that combine both rural and urban. This is one of the few things that is common to all African cities.

Roberts (1995) argues that there was evidence in Latin America of two important rural trends that influenced rural–urban migration. These two trends involve links between the cities to which migrants move and their rural origin locations. He argued that these trends were contradictory. The first is what he describes as the 'peasantisation' of the rural economy that retains the population on the land and creates strong rural–urban links. The second trend is of land consolidation and mechanisation, expelling the rural population to the cities where they have to seek employment. This is referred to as 'proletarianisation', as it creates proletarians (named after the lowest class of ancient Roman citizens) or low-income urban labouring classes of people who previously had survived on the land. Rural–urban links may well be affected if the local agrarian structure focuses on capital-intensive production which in developing countries is largely intended for export rather than for supplying the urban areas. Some systems may even retain rural labour, but the labour may be landless – rural proletarians – and, in some Latin American cases, may live in urban areas, migrating to rural areas when labour is in demand. By contrast, the peasant and subsistence form of agrarian structure retains relatively large numbers of people on relatively small amounts of land through intensive labour use. When the type of farming becomes more highly commercial, reducing labour input in favour of mechanisation, the Latin American evidence suggests that this results in high rates of out-migration to cities and often also abroad.

> Areas of peasant production present a complex set of economic and social relationships in which family enterprises span rural and urban locations, with town migrants retaining rights to land and aiming to return eventually. Peasant (subsistence) farmers not only survive in the present period but have increased in number since the 1940s, providing food for the large urban centres and a subsistence base for those working seasonally in commercial agriculture or even in urban employment.
>
> (Roberts, 1995: 97)

The report by the International Conference on Population and Development reflects concern at the process of urbanisation that is taking place when population data are considered at a global level (ICPD, 1994). For example, Potter and Lloyd-Evans (1998) report that during the 75-year period from 1950 to 2025, the global level of urbanisation will have risen from 29 per cent to 61 per cent. The year 2000 was considered significant in demographic terms because the world's population reached 6 billion for the first time. The lesser-known demographic statistic that was possibly as significant is that some estimates have put the world's urbanisation level as having reached 50 per cent for the first time as well. This puts us on track for 61 per cent global urban population by 2025. Figure 4.2 illustrates the UN's estimates of regional differences in urbanisation, while Figure 4.3 illustrates the variations in urban transition in a range of major regions.

In the developed countries, there is a population trend out of cities in search of pleasant rural living environments, known as 'counter-urbanisation'. Meanwhile in the developing world the contrasting process of urbanisation is prevalent. This means that much of the shift from rural to urban-based society will take place in the developing world. However, it is important to note that the level and rate of urbanisation will vary among countries. In Latin America and the Caribbean 75 per cent of the population live in urban areas, which is close to the average for Europe (UNFPA, 2003). By contrast, the *West Africa Long Term Perspective Survey* (Club du Sahel, 1994) showed that in 1960, West Africa's urban population was 14 per cent of the total. By 1994 it had risen to 40 per cent and was forecast to rise to 63 per cent by the year 2020. However, there are major differences between countries. Mali, Niger and Chad, for example, had urban populations of around 15–20 per cent in 1990, while Senegal and Nigeria had already reached 50 per cent. However, whatever the country, the increase will be steep, around

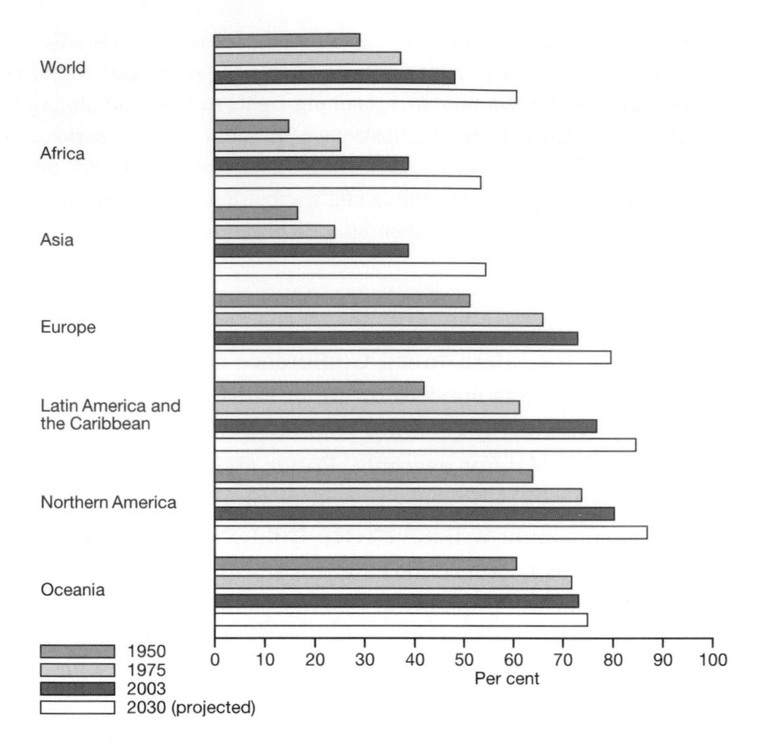

Figure 4.2 ***Regional levels of urbanisation.***

Source: UN (2003).

4–6 per cent per year between 2000 and 2020. It is expected that Africa and Asia will experience the fastest rates of urbanisation. For example *Cities in a Globalizing World* (UNCHS, 2001) reported that, while four of the top ten city populations are Asian and five American (three Latin American and two North American), Lagos has grown from a population of around 6 million in 1985 to 13.4 million, placing it sixth in the table in 2000. The report, however, also predicts that Lagos will continue to grow at a rapid rate, exceeding 20 million by 2010. So although there are global and regional trends, there is much difference and diversity and the pattern of rural–urban population distribution is far from homogeneous.

Just as the pattern of distribution is highly diverse, the empirical evidence suggests considerable difference in the pattern of movement between the urban and the rural realms. Theories of population movement in the developing world focus on rural-to-urban migration as this is by far the dominant flow.

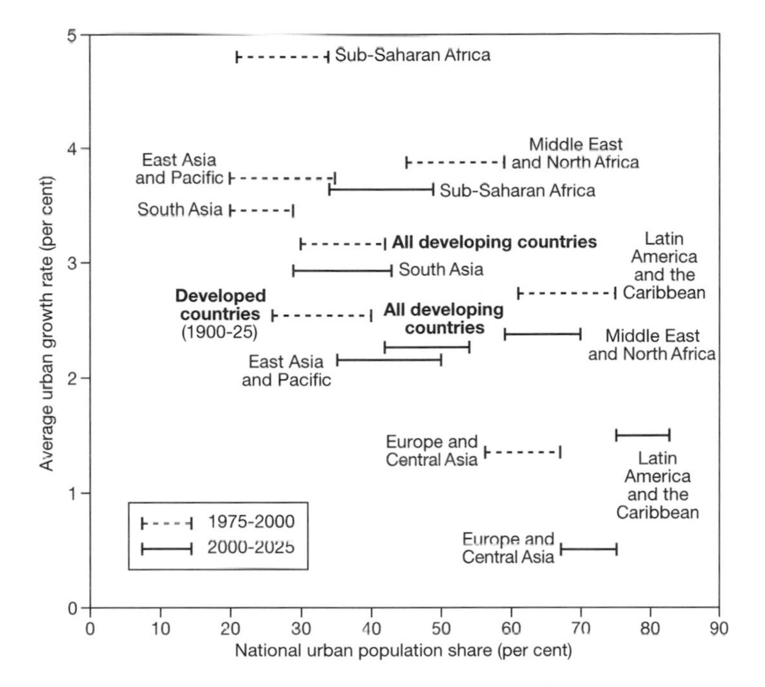

Figure 4.3 ***Regional comparisons of urban transition.***

Source: World Development Report 2003 by World Bank staff. Copyright 2002 by Oxford Univ Press Inc (US)(B). Reproduced with permission of Oxford Univ Press Inc (US)(B) in the format Textbook via Copyright Clearance Center.

Reasons for rural-to-urban migration

Some theories of the motivation for migration emphasise the importance of urban–rural differentials in incomes and resource allocation. This has been explained in terms of government policy and the differentials in development rates – particularly because industrialisation tends to favour urban areas (Drakakis-Smith, 1996). Todaro (2000) has developed a model of rural–urban migration that starts from the assumption that migration is an economic phenomenon. This means that in spite of high urban unemployment, expected income benefits from a move to the city are conceived as the main motivating factor in the decision to migrate. Todaro summarises the model as having four basic characteristics:

1. Migration is stimulated by rational economic considerations of relative benefits and costs, mostly financial but also psychological.
2. The decision to migrate depends upon an expectation of urban–rural wage differences, which is based on actual rural–urban wage differential and likelihood of obtaining employment.

3. The likelihood of obtaining a job is directly related to the urban employment rate.
4. Migration rates in excess of urban job opportunities are likely under these conditions. High unemployment rates are the inevitable outcomes of imbalances between rural and urban areas.

(Todaro, 2000: 309–310)

Todaro's policy recommendations, therefore, focus on the employment and income differentials between urban and rural areas. Such policies must take account of the fact that efforts to reduce urban unemployment may actually increase rural–urban migration, resulting in higher urban unemployment and lower agricultural outputs. Others argue that it is important to understand the variety of processes influencing the patterns of migration. There is empirical evidence to support the thesis that while the general development trends in Nigeria influenced rural-to-urban migrations, the particular circumstances in the country at certain times promoted such migration to a greater or lesser extent (Meagher, 1997). For example, the wide-scale Sahelian drought of 1972–74 forced many northern Nigerians, and those living in nearby countries, to flee to Nigerian cities in search of food and work. This resulted in increased permanent migration and increased cyclical migration. Since then additional droughts, low rainfall and poor harvests have helped maintain a high level of periodic migration.

While the effects of the drought were felt throughout the Sahelian zone, Nigeria also experienced an oil boom which, Meagher (1997) argues, increased the differentials in investment in the urban and rural areas. Increased levels of urban investment and increased urban wages were accompanied by neglect of the agricultural sector. The increased levels of wealth and spending in the cities led to increased demand in the informal sector. This led rural peasants in the far north of Nigeria to 'discover' Lagos, and the Yorubaland cities of south-eastern Nigeria as they sought opportunities of urban employment further afield (Mortimore, 1989). As Potter *et al.* (1999) report, there is evidence that this pattern of migration dates back earlier than the 1930s, when records indicate that migrants from northern Nigeria went to the Gold Coast (Ghana) in search of seasonal employment. This is an attempt by the mainly male migrants to find employment – labouring, fishing, petty trading and work in craft industries – bringing much needed cash supplies back to the home areas. However, the Hausa name given to this form of migration, *masu cin rani*, or 'men who while/eat away the dry season', indicates that a key part of their intention, and perhaps that of other seasonal migration systems, was to conserve supplies at home (Potter *et al.*, 1999). This is a clear example of a situation where rapid economic growth focused in the cities of Nigeria failed to 'trickle down' to the rural areas, leading the rural dwellers to move to the city in search of income. In some cases these pressures led to civil unrest and rioting in the cities.

However, during the 1980s, when the Nigerian economy went into crisis and the government introduced structural adjustment policies, the outcome adversely affected both urban and rural sectors, with a net shift in the rural–urban terms of trade in favour of rural areas. The increased economic pressure on urban households led to urban residents seeking alternative livelihood strategies. Meagher (1997) suggests that there is evidence that a breakdown in social networks between urban and rural areas during economic crises has meant that urban residents seek urban-based strategies for survival. For example, urban-based residents have established urban farming activities, either farming in cities to reduce their household food costs or managing farming in their rural home villages through relatives (see Chapter 2 above). Life in the rural areas was therefore perceived to be no better than life in the city.

Meanwhile, in the rural areas the high cost of inputs and expenses and the low returns from farming have served to increase the importance of non-farm employment, and this has often also led to cyclical migration. As economic pressures have mounted, both rural and urban residents have been less willing to add to their household costs by taking in migrant relatives. The resulting pattern has meant that those with wealth already and strong networks have tended to benefit under the conditions of structural adjustment while those without wealth lose out further. In northern Nigeria, those benefiting have tended to be southerners, in particular the Igbo from the south-east of the country (cf. Bryceson, 1999), who have been able to invest in their activities and use their strong links with commercial networks. By contrast rural dwellers have often used Koranic scholarship opportunities to stay with a master in town while working in petty jobs during the day. Meagher (1997) speculates that these contrasting social networks may have contributed to reinforcing the ethnic and religious tensions which have become exacerbated even since her paper was published.

One of the reasons why circulatory migration is so important may also be related to the issue of land tenure. Urban residents frequently return to their rural home in order to ensure that their claims to land are not lost. In the case of Zimbabwe, this has been identified as a major problem, resulting in the underutilisation of agricultural land (Potts, 1997). Such land is widely seen as a safety net and, though usually not constituting a considerable plot, is nonetheless an important source of security for urban migrants. At one stage the government proposed that non-farming urban dwellers should lose traditional rights to land. However, this policy was not implemented and would certainly have been unpopular. Potts's (1997) survey of urban residents indicated that they considered urban living to be risky and their access to and frequent returns to their rural 'home' represented an important component of their livelihood strategy.

Katz (1997) criticises the Todaro model because it treats migration as an individual and not a household decision. She argues that while it may be possible that people migrate as a consequence of expected income differentials, it is doubtful that they take these decisions on their own. Family or household factors may determine which member of the group migrates and ensure that the individual is supported so that the remaining members derive maximum benefit from the initiative. However, much of the focus of attention has been on the major cities. Parnwell (1993), for example, focuses on rural areas, arguing that village to city migration may act as a kind of safety valve, allowing people to get away from the grind and uncertainty of their rural livelihood.

Bryceson (1999) reports that the motivation behind migration can vary from one region to the next. In some countries there is evidence that rural–urban migration makes a limited impact on the economic livelihoods of the rural communities. She cites examples of research conducted in Ethiopia where soldiers returning from the war with superior education and skills found themselves no better off than men who had remained behind in the village. This may be to do with the fact that rural–urban linkages were minimal and returning migrants had few opportunities of capitalising on their advantage. Bryceson also reported some hesitation in taking advantage of rural–urban linkages among rural dwellers; in particular, she mentions rural women being disillusioned about the level of remittances being sent home. There is also some concern that the rapid spread of HIV/AIDS may affect those involved in forms of migration that take them away from their family home for long periods. Women in Africa see their husband's urban employment as important to household income, but there is worry about what the

Box 4.3

The migration experience in South Africa

During the apartheid era in South Africa, the government had in place a legal framework which prevented the settlement of the black population in cities. This was enforced by requiring them to carry a pass which indicated where they were from; only if black people had employment in a city were they allowed to be there. As a consequence, many black urban workers had very long commuting journeys (Smith, 1994). This kept the majority black population based largely in rural areas, highly mobile and highly vulnerable. It maintained labour costs at a low level and prevented the development of a large low-income urban population, concentrating urban management in the hands of the white elite. After the democratic elections in 1994 and the abolition of the Pass Laws, it was feared that huge rural-to-urban migration would result. For example, a South African newspaper, *The Saturday Star*, estimated that in the period immediately after the 1994 elections some 200,000 people moved into Gauteng Province and that 20,000 were moving in every month. Similar trends have been noted in migration to the Western Cape Province. This migration is presenting challenges to the government's early successes in tackling the massive housing shortages among the majority black urban population. By 1999 some 3 million people had been rehoused, the government building 680,000 houses for low-income residents and providing 959,000 housing subsidies (Lester *et al.*, 2000). Figure 4.4 illustrates the recent improvements in housing provision, showing the

Figure 4.4 ***Khayelitsha Township, Cape Town, South Africa.***

regularisation of plots and the larger buildings that are schools in the background, while in the foreground it is possible to see the provision of road networks and electricity, and investment by residents themselves in extending the basic structures by building walls and gates.

In a comprehensive survey of residents of informal housing areas in the province of Gauteng, Stevens and Rule (1999) found that 81 per cent had moved from elsewhere in the same part of Gauteng and 45 per cent had actually been born in Gauteng. They argue that this is set within a context of a number of other surveys conducted in the mid-1990s that found that the majority of migrants had moved from one settlement to another within Gauteng. The conclusion is that most residents are not newly arrived in-migrants.

In the same study Stevens and Rule carried out focus group discussion on the difference between urban and rural. They found that the main interest in urban life related to employment and income-earning opportunities:

> I love township life. Jobs are available in townships, unlike in rural areas. There are jobs available in clinics and shops. In other words, one cannot live a miserable life in townships. (Employed woman, 40+ years, Eatonside)
>
> In towns there are many piece jobs. If I am not working I can go looking for someone who can offer me some washing to do. In the rural areas you cannot, because we are all the same. In towns you can offer to clean someone's house and maybe raise yourself some R20. In the rural areas this is impossible. (Unemployed woman, 40+ years, Johandeo)
>
> (Stevens and Rule, 1999)

Meanwhile many of the urban residents surveyed had relatives including children living in rural areas, whom they visited regularly. This indicates the importance to them of the links to the rural areas and the importance of the links to those in urban settlements for their rural-based relatives. Such findings reinforce the theory that decisions to move, whether rural–urban or urban–rural, are largely based on increasing livelihood opportunities.

men are doing while away from home. Box 4.3 illustrates the implications of migration for individuals in the case of South Africa.

Implications of migration

It is important to remember that migration activities have implications for those moving, those left behind and those at the destination. Thus rural–urban migration has implications for both areas. De Haan (2000) argues that migration can result in inequality. If someone migrates, the resulting benefits and disadvantages are not always predictable and often depend on the nature of the relations between the migrant and those affected in the household or community. The outcome of a migration event is therefore influenced by the role of the migrant in social relations in both the urban destination and the rural origin. As McGee put it over thirty years ago:

> It is therefore necessary to avoid the assertion that geographical mobility – involving a move of population from the countryside to the city – is to be equated with social or economic mobility . . . simply an introduction to new types of economic activity and social relations.
> (1971: 29).

The change brought about by a transition from urban to rural will not necessarily bring about a change in social structure or growth in economic activity.

Katz's (1997) argument discussed earlier that migration may be a household decision implies that individual welfare losses may be involved, but the good of the family is the ultimate aim. Some researchers argue that this impacts most upon the lower-status members of the household such as women, who are likely to become what Potts (1997) describes as 'tied movers' and 'tied stayers', suggesting the individual is overruled in the decision to migrate or stay.

Gugler (1991, based on an earlier paper in 1969) argued that in eastern Nigeria urban dwellers considered themselves 'strangers to the city'. Migrants could all point to a rural place they called 'home' and many intended to retire there or at least be buried there. The result could be an urbanisation of rural places and a ruralisation of urban places as migrants move between the two, taking elements of their lifestyles with them.

It is important to stress that not all the colonial powers established and used their cities in exactly the same ways. Indeed, some researchers have argued that colonies were integrated into the global economy far earlier than they would have been if they had stayed independent, and that this brought early economic growth. An alternative approach emphasises the advantage of the relations for the metropolitan countries, arguing that the colonial period can best be characterised as one of exploitation (Frank, 1967; Harvey, 1973). During this period the colonial capital was the gateway through which power was exercised, and capital and goods were exported.

Post-independence analysis has similarly focused on the relationship that the primate city has had with its hinterland. Early in the post-colonial period the need to overcome the perceived urban bias was emphasised. Five strategies have been adopted based on this idea in different African countries to try to stem this influx of migrants:

- strict control over immigration to the towns or banning it altogether (the Republic of South Africa);
- forcible return to the rural regions (Mozambique);
- scattered urbanisation (Nigeria);
- decentralisation by promoting medium-size towns and by developing the regions (Zambia, Algeria);
- creating new capital cities (Nigeria, Tanzania).

Other countries throughout the world have attempted similar strategies. Box 4.4 illustrates China's approach to reducing population migration.

Thus, the question arises: does a dominant city facilitate development and economic growth by connecting the rest of the country to the opportunities of the global markets, or does it represent the post-colonial gateway through which multinational companies and Western governments continue to exploit the developing world? Or are both processes at work? This poses a dilemma that is well expressed in the ICPD report: 'The continued concentration of population in primate cities, and in mega-cities in particular, poses specific economic, social and environmental challenges for Governments. Yet large agglomerations also represent the most dynamic centres of economic and cultural activity in many countries' (ICPD, 1994: para. 9.12).

The Club du Sahel suggests in its study that the rapid growth of West Africa's cities over the next 20 years will result in the hinterland of each city expanding outwards, overlapping each other and even crossing international boundaries. Rather than this resulting in a crisis of urban sustainability, they suggest that it will raise the demand in the area and consequently improve the market performance and productivity of the region's rural areas (see Box 1.3). More recent reports based on the *West Africa Long Term Perspective Survey* (Club du Sahel, 1994) have therefore focused on how to make this possible.

Relations between town and countryside in West Africa could be hinged on:

Box 4.4

China's policies for controlling population movements

China's robust attempts to curb rapid population growth have been much discussed. However, the country has also experienced rapid urban population growth. Over the last 50 years, the urban population has grown in absolute terms from 57.7 million in 1949 to 351.7 million in 1995, representing a mean annual growth rate of 4.1 per cent per annum over almost a half century. Figure 4.5 illustrates the process of urbanisation as the population changes gradually from being 87.5 per cent rural based to less than 70.1 per cent rural based. Pacione (2001) suggests that much of the growth took place in the period after 1980 as government controls on migration and private markets in food and housing eased. This means that the urban population doubled on average every 17 years during the period 1949–95, but the growth is likely to have been far greater in the late 1970s and early 1980s. While the rate of growth seems to have slowed somewhat in recent years, the decade from 1985 to 1995 resulted in average annual growth of 3.6 per cent per annum or a doubling rate of 20 years.

In order to control the movement of the population in China and to create the kind of socialist society they envisaged, Mao Zedong's Chinese Communist Party decided that a complete reorganisation of the country would be required. Among other things, this involved the creation of a household registration system (*hukou*) and the commune system. By 1952 the commune system incorporated 40 per cent of the rural population and by 1957 it involved 95 per cent of the peasant population. The main change was a move from private land ownership to a communal system of labour organisation and collective management of the land and inputs. Each rural family had a private plot to meet its own household needs, but it also provided labour to its production team.

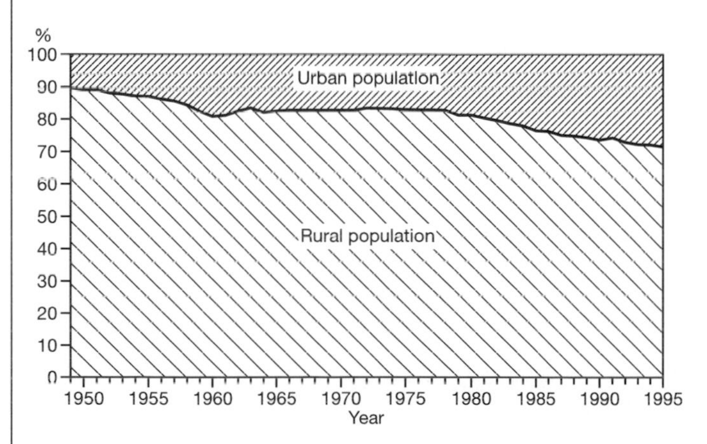

Figure 4.5 ***Urban and rural population of China 1949–95 (%).***

Source: adapted from Knight and Song (2000).

The *hukou* system, however, was more than just a registration system designed to identify personal status and provide population statistics. It was also intended to regulate population distribution. Chan and Zhang argue that *hukou* 'is one of the major tools of social control employed by the state' (1999: 819). *Hukou* classification determines rights in specified areas, for example supply of food, access to subsidised food and rights to employment. This meant that rural visitors without a local urban *hukou* required permission to be in an urban area and often had to bring food with them for their stay in an urban centre. Rural agricultural dwellers wishing to move to a city required permission from both their area of origin and their area of destination for reclassification from rural to urban *and* from agricultural to non-agricultural status. Such a change was subject to both quota and policy issues, the latter defining the kind of person eligible for conversion. The only way of obtaining food was by ration coupons, which were only made available to registered residents. Thus rural dwellers were discouraged from moving to cities because their access to rationed food would be restricted (see also Chapter 2 above). In addition, urban enterprises needed permission to employ labour (Knight and Song, 2000). While the *hukou* system remains in place, it has eased significantly since the 1980s. 'Spontaneous' migration to jobs in cities is tacitly accepted and temporary residence permits are easier to obtain for non-work reasons, such as study, visiting relatives, providing domestic service and receiving medical care (Chan and Zhang, 1999).

One of the key difficulties in any consideration of 'rural' and 'urban' relates to the statistics available as the data are collected and categorised according to the administrative definition in the state and at the time concerned. For example, Knight and Song (2000) report that in 1964, during the second census, a town was defined as consisting of more than 3,000 permanent residents, of whom at least 70 per cent were considered 'non-agricultural'. In 1984, a town was redefined to include all seats of

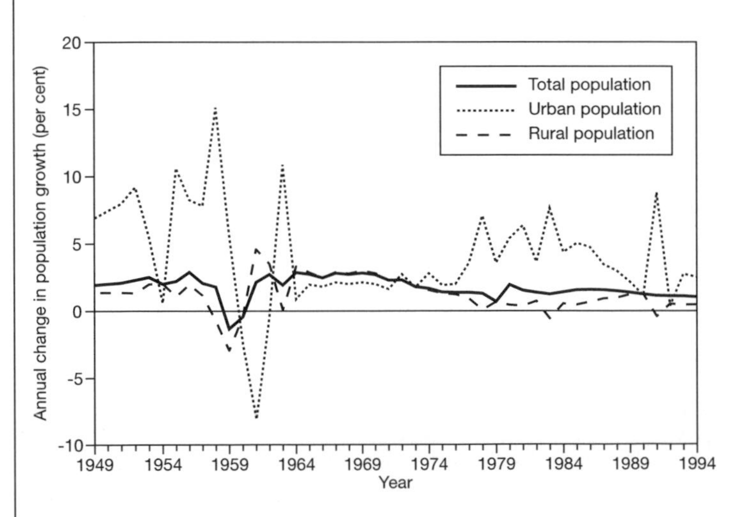

Figure 4.6 ***Annual percentage change in population growth in China (1949–94).***

Source: drawn from data in Knight and Song (2000).

county government and areas with at least 2,000 people considered 'non-agricultural'. This is further complicated by the fact that the censuses up to 1990 counted people in their place of normal residence defined by their *hukou* registration, rather than where they actually were (see also Box 1.2). Knight and Song (2000) argue that this is likely to mean that the data presented in Figure 4.6 underestimate the urban population. It also makes an accurate analysis of the trends difficult, as the categorisation of the data changed during the period of the time series. Figure 4.6 shows a pattern of change that is far from smooth, but it is difficult to tell whether jumps in the percentage change data are a result of administrative changes or actual demographic change.

> a large number of urban centers that will have strong economic ties based on exchanges of labor, goods and services with their surrounding rural areas. A network of local economic areas will emerge – unheard of thirty years ago – and just beginning in countries like Nigeria. These 'market watersheds' will cut across national borders in ways that will encourage economic growth.
>
> (Club du Sahel, 1994)

De Haan (2000) argues that the implications of migration can be negative as well as positive, the outcome depending on the social relations between the migrants and others affected (see Box 4.2). It is important therefore that governments are sensitive to the informal institutions that structure and facilitate migration processes. It is these institutions, he argues, that are the key to how migrants' decisions can support livelihoods. For example, it is possible to provide information about migration opportunities and consequences. This can, in turn, facilitate remittances and develop ways of enhancing the productive impact of migration decisions. De Haan (2000) argues therefore that consideration should be given to the geography that underlies many livelihood strategies such as migration. The way in which households and communities overcome the difficulty of distance implies that areas of origin and areas of destination should be considered as a single space for livelihood activities. Thus households that bridge the urban and rural divide are considered as one (see Box 4.3).

During times of economic difficulty a range of survival strategies can be adopted by those, whether urban or rural, who have access to urban and rural resources, for example household members located in both areas or moving between rural and urban bases ('fluid domesticity') (Potts, 1995). There is also evidence that net migration

Box 4.5

Voices of the migrants: the case of Durban

Some researchers focus on the motivations of rural–urban migrants. They have found that a range of complex motivations are cited, in addition to the usually foregrounded economic reasons. For example, Smit (1998) identified the following three main reasons for rural–urban migration in a study of low-income migrant households in Durban, South Africa.

Looking for work. Probably the most common reason for moving from a rural to an urban area is to look for work. The poverty of the remote rural areas of the former homeland of KwaZulu (now part of the province of KwaZulu-Natal) drives people to go to Durban in search of work. For example, in 1987 Victoria, who was then 22, left Mandini because there was no work there, and moved to Durban in search of employment. She left her older children behind with her mother, but took her baby with her to Durban where she got a job.

Political violence. The threat of political violence is a very real 'push' factor in parts of the world where there are political tensions. For example, the Ngcobo family left Ndwedwe in 1990 to move to Durban to live with people they knew. The reason they moved was that their homestead of two *rondavels* (round houses traditionally made from wattle and daub) had been burned down as a result of political violence.

Personal conflict. As well as political tension, personal tensions may prompt some to leave, particularly when rural populations are under considerable social and economic pressures. For example, Nicholas and Clara left their rural homestead in Ixopo to move to Durban in order to avoid problems caused by clashes between their families.

On arrival in Durban, the motivations for migration may influence the strength of ties to the rural areas. Smit (1998) reported that 39 per cent of his sample were members of what he describes as a multiple-home household, with homes in Durban as well as in the rural 'home'. The economically active members of the household are usually located in the urban home and the inactive members, such as elderly relatives and young children, in the rural home. Of Smit's sample, 49 per cent returned to their rural home every month. The significance of this is that the rural and urban homes are interdependent. The urban-based household members visit and send money to the rural household, while the rural household looks after the dependants, and provides a location to which the urban-based family can plan to retire or go for holidays and escape from the city life and violence that can affect low-income areas in Durban.

rates to cities slow down during the implementation of structural adjustment policies, as people move back to their rural origins because of economic restrictions.

Income-earning opportunities are the incentive that draws many rural inhabitants to the city. Other reasons include the 'push' of environmental degradation – either long term, such as land degradation, or catastrophic, such as a major drought or flood. Box 4.5 provides individual cases of motivations for moves in the case of people migrating to Durban in South Africa. Some researchers argue that many national education systems are urban-oriented and encourage young rural people to go in search of urban-based jobs that will make use of their skills and pay them more than farming or other rural occupations.

Much of this research underpinned the policies that were put in place to redress the rural–urban balance. Most such policies focused on population redistribution measures aimed at reducing rural-to-urban migration, in particular migration to the larger cities. These include land resettlement and agrarian reform, resettling rural populations into sparsely populated areas or on to newly acquired public land. Examples of such policies include Ethiopia's resettlement of people into the regions bordering Somalia in order to reinforce its claim to disputed areas. In Brazil, people were encouraged to resettle in the Amazon basin as part of a programme of opening up the area by major trunk road construction. In Indonesia, over 6 million people were moved from the densely populated island of Java to other regions of the country, including, most controversially, Irian Jaya and East Timor. Irian Jaya is the western half of the island of New Guinea, which Indonesia incorporated in 1969, removing land from the local population for the Javan in-migrants. A similar series of events occurred in East Timor, which the Indonesian army annexed in 1975 (Rotgé, 2000). However, since East Timor, now known as Timor Leste, gained its independence in 2002 many of the Javanese Indonesians have left (Macaulay, 2003). An alternative policy attracts population movements to medium and small towns by establishing them as growth centres and providing improved housing and infrastructure. Examples include the *agrovilles* of Pakistan, new towns in Malaysia and the rural centres and intensive development zones in Zambia (Hinderdink and Titus, 1998). In large part these have been unsuccessful. Hinderdink and Titus suggest that the reasons for their failure include the following:

- The main economic urban linkages are focused on international trade.
- The rural centres were often focused on urban industrial employment, and as such unsuccessfully attempted to attract investment from the urban core.
- The growth centres are seen as 'vanguards of exploitation' by some researchers, facilitating urban exploitation of rural areas.

Atkinson (2000) provides evidence that the majority of rural-to-urban migration recently recorded in the Philippines is in fact to the smaller towns rather than the larger conurbations such as Manila. Similarly, Potts (1997) finds that the rural–urban migration patterns in Africa are highly complex. She identifies variations between countries and according to changing economic circumstances within one country. For example, pressure for rural land in Egypt and parts of Kenya, resulting in commercialised rural land markets, may have constrained urban-to-rural migration as a crisis strategy for urban residents. However, where rural land is more easily available, as in Tanzania or parts of Ghana, it may offer an 'escape route' for urban residents made redundant from their employment. Even if urban dwellers are not made redundant, they may use agricultural activities in their rural 'home' areas or in peri-urban areas as a way of keeping costs low or earning additional income.

The inexorable growth of the world's urban population is a cause for concern in the countries most affected by it. Some governments have therefore tried to reduce rural–urban migration through a range of policies including transmigration, migration controls (as in China, see Box 4.4, or South Africa, see Box 4.3), de-urbanisation and industrial dispersal. These policies have often been criticised for not considering human rights or the environmental implications of locating industries in rural areas. For example, Indonesia had a transmigration programme which was intended to stop people moving to urban areas by resettling them in less populated areas; thus people were moved from the island of Java where a large proportion of the population of Indonesia is located, despite the country comprising around 14,000 islands (Atkinson, 2000). Most migrants benefited, but perhaps at the expense of indigenous peoples and of the environment, notable impacts being the destruction of rainforests (UNESCAP, 2001) and the suppression of the independence aspirations of the population of Timor Leste (discussed earlier).

One of the most radical and brutal state approaches to the removal of rural–urban disparities was made by the Khmer Rouge in Cambodia. They evacuated the urban areas in order supposedly to build a new egalitarian agrarian society. However, their anti-urban ideology resulted in the destruction of technology and the elimination of any benefits that cities might bring even to an agrarian society. Many people died during the draconian evacuation policies and the harsh life in the rural areas. After the Khmer Rouge fell in 1979, there was a huge influx of rural migrants into cities and towns in search of security and employment.

Conclusion

There is a clear pattern of population growth in most cities in the developing world. The evidence discussed in this chapter suggests that this is a result partly of natural growth – people being born in cities – and partly in-migration from rural areas. The UNFPA (1996) suggests that the relative contribution of each to urban growth is equal. However, beyond this, the picture becomes complex. There is evidence to suggest that migration between urban and rural areas involves large numbers and can go in both directions. In some cases, circulatory migration is a key part of both household and individual livelihood strategies. Some households adopt a strategy ensuring they have access to both urban and rural locations in order to take advantage of the complementary livelihood opportunities in each. For example, urban living appears to focus on social networks which provide opportunities for market-based activities, while kin and family ties – although breaking down in some locations – still constitute important rural social networks. There is evidence to suggest that under certain conditions – in particular conditions of economic adversity – the net flow of migrants may turn from rural to urban to urban to rural, as city dwellers flee adverse circumstances in search of a rural safety net.

Discussion questions

- Review the different forms of migration that are described in the literature on rural–urban population movements.
- Critically assess the role of the state in the management of migration between cities and the countryside in the developing nations.

- Review the main motivations for migrating from the rural areas to the city. Review the evidence for motivation to move from the city to the countryside.

Suggested reading

de Haan, A. (1999) Livelihoods and poverty: the role of migration – a critical review of the migration literature. *Journal of Development Studies* 36 (2), 1–47.

Potts, D. (1995) Shall we go home? Increasing urban poverty in African cities. *Geographical Journal* 161 (3), 245–264.

Zhu, Y. (2000) In situ urbanization in rural China: case studies from Fujian Province. *Development and Change* 31 (2), 413–434.

5 Ideas

Summary

- **The development of information and communications technologies provides powerful potential for countries and regions affected by poverty to overcome their challenges.**
- **One main means by which improved communications can improve the possibilities is by connecting rural and urban areas more effectively and at less cost.**
- **However, there is concern that the forecast benefits of new technologies have been over-estimated. This chapter illustrates that lessons can be learned about the relations between urban and rural areas by examining the impacts of more traditional media, such as print, radio and television.**
- **Finally, the chapter addresses the role of media in the construction of urban and rural identities and livelihood strategies. This shows that the introduction of improved communications between urban and rural areas can be detrimental as well as positive to rural areas, challenging traditional identities, creating unrealistic ideas of urban life and reinforcing existing power relations, rather than increasing the access of marginalised rural groups to economic, social and political capital.**

Introduction

This chapter sets out to consider the flows and influence of ideas on the relationships between rural and urban places in developing countries. This will involve a brief discussion of the theoretical issues, in particular the popular constructions of urban and rural and the decline of the rural peasant and its corollary, the emergence of the urban middle class. The chapter will then consider the information and communications sectors, including mass-market media, such as television and radio, and reciprocal communications, in particular telecommunications.

This chapter focuses on two aspects of ideas. The first relates to the information 'architecture' of developing areas and how the urban and rural realms are 'mapped' on to the communications infrastructure. For example, to what extent are rural places 'present' in cyberspace and how does this impact on the flow of ideas between rural and urban places? This sounds a little futuristic and perhaps not relevant to the rural areas of the developing world. However, some people are trying to make the technology of the internet available to low-income rural dwellers so as to give them the opportunity to 'leap-frog' into the twenty-first century. Indeed, during fieldwork the author was told an anecdote that illustrates the potential power of this technology (Ntiro, personal communication). A small internet café was established in a town in southern Tanzania. Two cashew-nut farmers interested in finding out what the internet was discovered that they could access world cashew-nut price data. With this information, they were able to negotiate an improved price with their buyer when the next harvest season came round. This illustrates how a relatively low-cost and relatively short session on the internet can put the power of information in the hands of ordinary people. If such information which is in the 'public domain' (i.e. anyone who has access to the technology can obtain it) can empower cashew-nut farmers, what would be the effects of providing access more widely? Websites already exist that make available maps, remotely sensed images, international commodity prices and information on specialist markets that could genuinely overcome the friction of distance. However, to what extent is this aspiration plausible or possible? The Tanzanian cashew-nut farmers show a beginning. However, this chapter also demonstrates that considerable distance apparently needs to be covered before internet technology is widely accessible and widely beneficial. Some writers refer to this distance as the 'last half mile', emphasising the difficulty of connecting relatively close areas. However, others have described it as the 'first half-mile', perceiving the distance from the perspective of the potential users rather than the potential providers. This point is considered in greater detail later in the chapter.

The second aspect is the impact of ideas as they interact with places. For example, how do rural ideas affect urban places and vice versa, and to what extent is this interaction important or significant?

Rural–urban flows of ideas

While flows of goods and people are tangible and observable, the flows of information and ideas that usually accompany them are far more complex and difficult to research. This is partly because, although less tangible, the relations that structure the flows of other elements also influence the flows of ideas. However, since information and ideas can be produced anywhere, and modified and imbued with power and authority or weakened and marginalised, they are far more difficult to control or influence. This chapter sets out to discuss flows of ideas, information and ideology. However, partly because of the complexity of these issues much of the research focuses on the visible forms of transfers of information and ideas between cities and their rural hinterland, that is on the information infrastructure. Much of this chapter therefore examines how the media, information and communications technologies (ICTs) and networks of people and institutions act as conduits for flows of information.

As discussed in Chapter 4 above, the movement of people to and from cities and the numerical growth of city dwellers have had profound effects on societies more generally. In particular, increased movement of people has resulted in increased flows of information. Improved transport and communications mean that rural and urban areas are linked more closely than ever before. Such improved links bring the city to the countryside and the countryside to the city. Bryceson (1999) suggests that in sub-Saharan Africa this has prompted rural dwellers to aspire to an urban life with improved employment opportunities, while elderly urban dwellers look forward to returning 'home' to their village. She describes these people as 'betwixt and between': in one place, but yearning for the other. This mixed-location dweller can take a number of forms. For example, Schilderman (2002) argued that the links between the small town and the country are stronger because of the proximity of one to the other. His empirical evidence from a study in Peru suggested that, when possible, people moved to towns for the improved services and educational opportunities for their children, but often found their skills inappropriate for making an income.

This necessitated frequent return visits to participate in agricultural activities. This state of 'permanent circular migration' was intended to increase appropriate livelihood opportunities, but also had the effect of increasing their information networks. This social proximity of the urban and rural is also described by Dayaratne and Samarawickrama (2001) who emphasise the high esteem in which rural living is held throughout Sri Lanka:

> Sri Lankans . . . are proud to be seen as a 'peasant community' and continue to derive their value systems from their associations with agriculture. In fact, in the popular social stratification system of castes that has and continues to play a significant role in the community, the *govi* caste representing agriculture are considered the most superior and likened to castes of kings. Rural affinity is a characteristic to celebrate and has a significant influence upon the transformation of the rural space and communities.
>
> (Dayaratne and Samarawickrama, 2001: 2)

Sri Lankans' strong identification with agriculture and with rural society is likely to be at least in part a result of the majority of non-rural dwellers having been brought up in rural areas. One of the features of very rapid urbanisation is that a majority of urban residents may originate from rural areas.

In addition to general issues of difference in the way urban and rural are considered, there can be gender implications. For example, in his study, Schilderman (2002) found gender differences in access to information:

- The information needs of women in poor settlements are quite different from those of men, and are often ignored.
- Women's social networks are different from those of men. They are based much more around the neighbourhood and are generally with other women.
- Women's access to information from sources outside their social networks and communities is often restricted by men. However, such information can prove to be powerful and very effective.
- Women are constrained in many ways in accessing information. The reasons for this include their position within society, high rate of illiteracy and lack of authority. But at times they have managed to overcome these constraints, and in Sri Lanka several have become key informants as a result.

Clearly, the relative differences in the role and power of women in urban and rural settings result in a difference in the challenges they face in each setting.

The UN Commission on Human Settlements (also known as Habitat) promotes the importance of rural–urban linkages (see Table 1.1). It argues that nowhere are these more important than in the flows of ideas between rural and urban areas. These flows are often characterised as part of the urbanisation of the rural environment. For example, Critchfield (1994) highlights the power of urban and Western ideology over the rural hinterland in Asia, Africa and Latin America. He writes about watching television in a Punjab village in India: 'In Gungrali we saw how Nestlé's Maggi 2-Minute Noodles were catching on, thanks in part to the enormous popularity of the Hindi soap opera, *Hum Log*, which it sponsored' (1994: 439). This power is exercised through various urban-based mass media, such as newspapers, television and radio. This is compounded by the lower levels of access to the reciprocal and more horizontal systems of communication such as telephones and e-mail among the rural population. The International Telecommunications Union (2001) in its deliberations about more inclusive use of telecommunications technology makes a distinction between *conduit* and *content*. The contents of media communication are largely created by urban-based participants, and the conduits are largely located in urban centres with the same reach into rural areas. The result is an urban-dominated outlook, portraying urban lifestyles.

Chapter 4 above focused on the movement of people between cities and rural areas. There is some overlap with this chapter, because people can also act as conduits of ideas (Englund, 2002) and add to the credibility and power of ideas. In analysing what influences rural dwellers to move to the city, de Haan (1999) identified recent urban in-migrants as being partly responsible for transmitting a positive account of their urban experiences back to their rural counterparts. They may do so even if the reality of their experience is more mundane, in order to save face with rural friends and relatives. In addition, one of the key aspects of the survival strategy identified by both de Haan (1999) and Englund (2002) is the fact that they are able to build social capital in both the urban and the rural environment. The construction of social capital is therefore a key aspect of the flow of ideas.

Concern has been expressed in the past that national education policies are often developed by an urban-based elite. The resulting curriculum inculcates its urban ideology in school-leavers. At the most basic level, educational curricula emphasise reading, writing and arithmetic, usually in the dominant language of the country's government which is often the one more frequently used in the cities than in the countryside. There is evidence that urban-based media – whether the source is domestic or foreign – and education systems have an important influence on promoting rural–urban migration. Chambers (1982) also suggests that research into rural poverty has an element of urban bias: researchers from cities start their work in the ministries, government agencies and research institutes in the cities before going to speak to the rural poor, the subjects and intended beneficiaries of their research.

The construction of urban and rural identities is largely based on the various ways in which those living in urban and rural contexts gain access to information about the opportunities and threats of these realms. One key aspect of the construction of these identities (considered in the next section) is the rise of urbanism and the demise of the peasant. Another aspect is the role of the mass media in communicating information and ideas about urban and rural places and in implying the inequalities in the power and importance of these realms. A later section in this chapter therefore considers these media in more detail. Finally, a number of agencies are increasingly focusing on the role that reciprocal communication can play in the advancement of the poor in both urban and rural areas. So this chapter concludes with a discussion of the means by which this may take place, in particular focusing on the use of information and communications technology to facilitate the links between rural and urban areas.

The disappearance of the rural peasant?

Some researchers argue that peasants are disappearing from Africa (Bryceson, 1999) and South-East Asia (Rigg, 1998a, 1998b; Elson 1997; see also Box 5.1). Their disappearance appears to have different implications in different circumstances. In Africa Bryceson (1999) finds evidence that the 'de-agrarianisation' of rural areas is a response to the pressures placed on rural peasants. They respond by engaging in alternative livelihood strategies, some of which involve interaction with urban areas, in an attempt to meet their needs through income-earning opportunities. In Africa, subsistence production is an important backstop for both rural and urban dwellers. Access to land provides a livelihood asset that can be used by households even in urban areas, if the economic activities there do not provide sufficient income. Consequently, although there is evidence that households are engaging in strategies that involve inhabiting space in both urban and rural areas, many social networks appear to have a rural focus. Urban dwellers can thereby ensure continued security of tenure over their backstop subsistence plot. This 'subsistence fallback' (Bryceson, 1999), however, often fails due to an increasing mismatch in population and agricultural productivity or poor levels of investment. By contrast, in the highly commercialised rural areas of South-East Asia, the response is far more international and urban focused. The people still produce rice, but they produce more and faster and sell most of it. They buy most of what they consume; they move around, taking advantage of opportunities communicated to them through the mass media; their culture is more influenced by the city. Bryceson (1999) identifies four opposing tensions with which peasants are confronted:

- non-agricultural market experimentation versus reliance on an agricultural subsistence fallback in the quest for economic survival;
- household solidarity versus individual autonomy in the course of mobilising resources and social networks;
- social identity conflict arising from agrarian community conservatism as opposed to individual scepticism; and
- the strengthening or weakening economic foundation of rural livelihoods related to linkages between agricultural and non-agricultural activities.

(Bryceson, 1999: 27)

Under pressure, cities and towns offer opportunities to resolve these tensions through employment, market outlets, and members of kin or social networks who may be able to assist. In addition, in some cases in Africa there is evidence that rural communities are strengthening social ties and maintain and strengthen social links between urban and rural dwellers through the practice of ritual ceremonies. They thereby emphasise the positive importance of traditional and social ties and ensure that urban-based relatives are reminded of their roots and responsibilities. In South Africa, for example, Bryceson (1999)

Box 5.1

A farewell to Asian farms?

Jonathan Rigg (1998a) tracked the 'poor' in two rural villages in north-eastern Thailand between 1982 and 1994. He found that the average household found its income increased by 44 per cent, which is in line with the national trends during that period. However, the income differentials there did not widen, unlike the evidence for the nation as a whole. Rigg explains the widening of the national differential between poverty and wealth as largely the result of regional differences, and particularly differences between urban and rural areas. This said, however, Rigg's field observations indicate extensive penetration of the villages by consumerism, with luxury goods being far more prevalent in 1994 than in 1982.

In another village in northern Thailand, there is evidence of diversification of rural livelihood activities between 1974 and 1991. The evidence shows a shift from a dominance of farming (the main activity of 52 per cent of the respondents surveyed) to a dominance of wage labour, farming having declined to 4.8 per cent (farming as the main activity) compared to 51 per cent for wage labour (Rigg, 1998b).

Village dwellers in Vietnam have also reported that having a family member working outside the village was 'a *matter of survival*' (Rigg, 1998b: 502, emphasis in original) for the household. This transition from agricultural to rural economies increasingly dependent on non-farm economic activities is taking place across South-East Asia among wealthy and poor rural dwellers alike. However, Rigg identifies a differential in take-up of non-farm employment between wealthy and poor farmers. For example, the evidence suggests that wealthy farming households engage in more highly capitalised and better-paying activities, such as wholesale trading or taxi or bus ownership. Meanwhile the low-income rural dwellers are involved in easy-entry activities such as petty trading, basket making or unskilled construction work. Finally he commented: 'If the identification of the rural poor has less and less to do with agriculture and land, and more to do with nonfarm activities then many of the antidotes to rural poverty would also seem to lie outside the agricultural sector' (Rigg, 1998b: 518).

reports that in villages in Mooiplas the elderly women are emerging as the keepers of cultural tradition and organisers of rituals, thus reinforcing matrilineal ties and attempting to increase their authority and ensure their material security. Similarly, the author has met groups of elderly women in townships in the Eastern and Western Cape who are breathing life back into traditions, in part in response to tourist interest but also in order to remind relatives of their traditions and backgrounds. This places them in important positions as keepers of the culture and may provide an important source of income. By contrast, some areas see the incursions of urban influences as bringing many negative social consequences, linking increased contact with cities to increases in crime, conflict and land disputes.

The rise of the urban middle class

Parallel with the decline of the peasant in rural areas, and the development of permanent urban areas, is a phenomenon which may be influential on ideas about the role of cities in developing countries. It may also be important to the future role and therefore sustainability of these cities. This idea is what Forbes (1996) described as the growth of the urban middle class. This, he argues, is already in evidence in the changing make-up of the South-East Asian city. It is apparent in the changing structure of urban consumption, the rise of shopping centres, the development of condominiums, and the growth of South-East Asia as a source of international tourism, especially to Australia. This, in addition to the growth in access to media, will inform this growing social group about alternative political and social structures as they engage with other societies. Not least of these involve gender and lifestyle.

The influence of such changes is focused on urban rather than rural areas. The evidence suggests that the cities of South-East Asia, such as Singapore, Bangkok, Kuala Lumpur and Jakarta, will have an increasingly outward focus as they chase the markets for internationally traded services. This is likely to focus their relationship with rural areas on the ability of these to provide resources, in particular skilled labour. This export-orientated urbanisation will continue to attract rural–urban migrants who are less able to take advantage of such developments and lead to these cities continuing to accommodate wealth and poverty side by side.

The importance of the rise of the urban middle class is also to be seen in the influence that this social group has on the mass media, discussed above. As the more commercialised forms of mass media, in particular television, respond to the growing consumer market of this emergent middle class, concern is evident (in the discussion above) that this will affect the cultural role of such media and modify the content of the message they broadcast. Thus, the messages broadcast through commercialised mass media tend increasingly to reinforce the lifestyles of the middle class. It is therefore important that rural–urban linkages using mass media are supplemented with other forms of media, and that rural user groups have access to programme makers in order to ensure that these media are not simply broadcasting an essentially urban message to the whole country. More reciprocal communications systems, community broadcast systems and rural pressure groups can balance the urban dominance of many national broadcast media and ensure that the information and knowledge disseminated are relevant to rural as well as urban communities. Such bottom-up initiatives can also be assisted by appropriate policy contexts. For example, broadcasting watchdogs and ministries can ensure that there is a policy framework that supports balanced diffusion of information and communications technology and therefore balanced information and knowledge for development.

The spread of mass media

The spread of mass media has resulted in a further blurring of the difference between town and countryside. This has been facilitated by improvements in communications technology and by increased mobility which allows migrants to experience other places and report on them to their friends and relatives. Such communication may be based on the social networks that migrants use to facilitate their moves to and from towns, cities and rural areas (see Chapter 4 above).

Roberts (1995) reports that by the 1980s most Latin American citizens had access – albeit variable – to quite sophisticated media in the form of newspapers, radio and television. In many cases, despite relatively authoritarian state control, the information broadcast in these media communicated important concepts over which the states apparatus had little control. For example, Roberts argues that the media

> are often controlled by governing elites, as in the case of *Televisa* in Mexico, but their influence on public opinion is broader than their political propaganda since it includes images of lifestyles and family practices, national and international, conveyed through *telenovelas* (soap operas) that are avidly watched by an estimated half or more of the region's population.
>
> (1995: 201)

The presence of the international media in regions and events of conflict can provide a means of protection and independent coverage. Opinion polls have become routine in even the most authoritarian countries in Latin America (Roberts, 1995).

Radio

Radio has long been a key broadcast medium that has been widely adopted across the developing world. In particular, large international audiences exist for the worldwide broadcasts of radio stations, such as the BBC World Service, Voice of America, Radio Deutschewelle, and regional radio stations such as SABC and Radio Tanzania aimed at listeners beyond their own borders.

More recently improvements in the affordability and accessibility of the technology for good-quality broadcasts have resulted in increases in the generation of local content for radio broadcasts. For example, the development of small-scale broadcasting equipment has resulted in a proliferation of FM radio stations. In Mali in 1999, for example, there were 107 licences and 92 operational radio stations. With the development of low-cost, low-power radio station equipment, community radio stations are expected to proliferate in a number of countries. Kleih *et al.* (1999) found estimates for the cost of initially setting up a community radio station to range between £15,000 ($24,000) for a small station and £50,000 ($80,000) for a larger station. This would include the cost of broadcast equipment, transmitters, studio, vehicles and training of personnel. For the listener the cost is relatively low and can be made lower by sharing the resources of the radio and the cost of batteries. Figure 5.1 illustrates this, portraying a group of men who regularly meet in the village to listen to a radio belonging to one of them. They also provide a market for the informal cigarette seller. Radios playing at a loud volume are often used as a strategy by informal stall-holders throughout urban and rural areas in order to attract business.

However, there are reservations about the role that broadcast media such as radio can play in society as a whole. For example, the World Bank (2002) reported that in a sample of 97 countries 72 per cent of the top five radio stations were government owned. This contrasted with 60 per cent in television and 29 per cent of newspapers. According to the World Bank, monopoly ownership of broadcast media, whether government or private ownership, can restrict the reach of the media and also influence the information reported.

Figure 5.1 ***Group of men in a northern Nigerian village gather to listen to the radio.***

O'Farrell *et al.* (2000) argue that there is a large unmet demand for radio broadcasting. They illustrate the importance of such accessible media with some statistics. It is estimated that one person in ten people in the developing world has a radio. However, perhaps more significantly, many more have regular access to a radio even if they do not own one. For example, in Bangladesh, 23 per cent of males and 21 per cent of females own radios, but 71 per cent of males and 44 per cent of females have regular access to one. Thus many people listen to other people's radios or listen to them in public places. This suggests that radio is a medium with an enormous potential reach in countries like Bangladesh. It also suggests that using comparative penetration data, such as number of radios per thousand population or percentage ownership, is inappropriate. This is in contrast to the advanced-economy countries where many people may own several radios. Access to and use of radios is clearly substantially different in different societies. Radio may hold a more significant role in rural than urban areas of the developing world, as illustrated by Figure 5.1. O'Farrell *et al.* (2000) also point out that radio may be well suited to rural areas of the developing world, because many of their societies have oral-based traditions, meaning that men and women, young and old, rich and poor are all experienced in the generation, gathering, exchange and deciphering of oral information.

Television

Television is now a truly global medium. Potter and Lloyd-Evans (1998) show that ownership of televisions and radio receivers in the Caribbean is now near-universal. Even in low-income households in Barbados, ownership of video recorders was approximately 43 per cent. There is, however, concern about the televising of North American soap operas, which may lead to a mismatch between lifestyles and aspirations. Such phenomena may have contributed to what McGee (1991) describes as the diffusion of urban influence into the rural hinterland (see the discussion in Chapter 1 above). For example, Ping *et al.* (1999), researching rural-to-urban migration from households in Jiangsu, Anhui, Sichuan and Gansu provinces in China, were struck by the high level of television ownership in the rural areas of these provinces. They recorded a mean of one set per 1.17 households. Of those who lived in the village, 64.01 per cent said they watched television for one or two hours or more per day and 20.92 per cent watched for longer periods. Ping *et al.* (1999) suggest therefore that television plays a significant role in informing villagers about cities, especially the major cities such as Beijing, Shanghai and Guangzhou, through news programmes, and about Hong Kong through news and drama series located there.

While the spread of ownership of televisions continues there is also concern about the fact that most of the content generated is imported, in particular from dominant Western cultures. For example, much is written about the spread of channels such as CNN and MTV (though little yet has been written about the BBC World Service!), and this is linked to concern about cultural imperialism. This concern is also associated with the spread of global brand names such as Coca-Cola (leading to 'Coca-cola-isation') and McDonald's (leading to 'McDonaldisation', but also ironically the more serious use by the *Economist* magazine of the price of the 'Big Mac' hamburger as an indicator of the cost of living in different countries). However, Sreberny-Mohammadi (2002) argues that while many Third World governments have set limits on the importation of television programming, most multinational television programme providers are more concerned about the restrictions in the European Union market, where well-developed media businesses are competing for a large market with high disposable income. By contrast, Third World countries are problematic for most multinationals in this sector. Their market combines a low level of income and a high level of piracy. She also suggests that the problem of video piracy may be undermining the development of embryonic television production in the countries concerned. The Brazilian-produced television soap operas, called *telenovelas*, show signs of greater levels of commercialisation than equivalent American imports. This has led to concerns that there is a hybridisation and 'creolisation' of local television forms. The debate focuses on whether such popular forms as *telenovelas*, not only in Brazil but throughout Latin America, are a dilution of local culture or a local reworking of imported culture. Similar debates exist in Iran where, according to Sreberny-Mohammadi (2002), pirated versions of *Maximum Overdrive* and *Robocop* and of music by Madonna and Michael Jackson are popular. She also reports that in Asia *Rambo* has been remade, and remote Guatemalan Indians listen to old tapes of the Beatles.

Liebes and Katz (1989) carried out a multinational study of viewers of the 1980s soap opera *Dallas*. They argue that their results represent diverse readings of the narratives, rooted in the variety of different cultural backgrounds of the viewers. In essence, the received message of the soap opera is in part constructed by the viewers and their social and political context. For example, Russian viewers were more highly sensitised to the 'politics' of *Dallas* as a representation of the capitalist West, while Arab groups had a high sensitivity to the 'dangers' of Western culture. In contrast, the US viewers saw the programme more in terms of ideology-free entertainment.

The concern about global cultural imperialism often focuses on the volume of sales and the dominance of the US industry at the expense of others. For example, Clark *et al.* (2002) report that while the US accounts for 75 per cent of television exports, 80 per cent of this goes to seven countries. Meanwhile, regional markets have emerged that include the Francophone countries (France and its former colonies), Latin America, the Arabic world market, the Chinese market and a South Asian market. These media spaces do not have contiguous boundaries, but involve diaspora populations distributed across the world. Most notable of these is the UK Indian diaspora and the Latin American diaspora located in the US. Thus, while there is a geographical focus for the production of television media, the places where television is viewed are embedded in cultural contexts which may overlay different cultural interpretations and meanings. Wilk (2002) identifies three categories of viewer:

1 *dominated* viewers, who place no distance between themselves and the programme to the extent that they may fuse the programme with reality;
2 *negotiated* viewers, who accept the meaning of the programme, but interpret and adapt its significance to their own context;
3 *oppositional* viewers, who have critical distance from the programme, allowing judgements to be made about its truth, the motives of the makers and the impact of the medium.

The question then is how to map these viewing stances. 'A proponent of cultural imperialism can argue that the viewers in the metropole have critical distance while those on the periphery are dominated and deceived' (Wilk, 2002: 290). In his empirical work in Belize, Wilk (2002) found a profound ambivalence about television at every level in society, reflected in interviews with Belizeans and in newspaper clippings on the debate about television.

> The rural working class tended to criticise the moral content of programs and the direct effect of viewing on children, the middle class is more likely to see negative *social* and *cultural* effects. . . . The middle class has had television longer and, unlike the poor, for whom viewing is a social event, they now complain that television decreases social interaction and isolates people.
>
> (Wilk, 2002: 293; emphasis in original).

According to interviews with rural Belizeans, television appears to be an additional external factor that is affecting society, in particular family and cultural life.

Mankekar (2000) found that the regulation of television broadcasting in India was originally highly controlled, but has gradually been relaxed as it spread outwards from Delhi to other urban centres such as Mumbai (Bombay) in 1972, Srinagar and Amritsar in 1973 and Kolkata (Calcutta), Chennai (Madras) and Lucknow in 1975. As this relaxation took place, sponsorship and advertising began to take an important role in broadcasting. In a search for new forms of television compatible with its aims of state building, the state has used the Latin American model of *telenovelas* to present *teleserials*, and produced an Indian hybridised form of education through entertainment. They are produced in the metropolitan centres, though some are in regional languages. They and their associated sponsorship and advertising appeal (as mentioned in the case of *Hum Log*) to the upwardly mobile middle classes represent an attempt to draw them into the project of the construction of a national culture. The result is that 75 per cent of the Indian population is 'covered', meaning that they have access to television, whether this is community viewing in rural villages, extended household viewing in low-income urban areas or family viewing in middle-class households. This results in a very varied definition of the activity of 'viewing', as many women, for example, may have limited power over the decision of what to watch. For some Indian women, 'viewing' was done by turning up the volume of the television so high that they could hear the soundtrack while working in the kitchen or washing clothes on the veranda (Mankekar, 2000).

Abu-Lughod (2000) argues that one effect of television in Egypt is to give women, the young and the rural as much access to stories of other worlds as urban men already have. However, she also warns that television can turn these people into subaltern objects in need of enlightenment, or subordinates open to influence and domination. But, she argues, 'from my ethnographic work in the village, I would suggest that the villagers make elusive targets for the cultural elite's modernising message . . . [the reason is] the very ease with which they have incorporated television into their everyday lives' (Abu-Lughod, 2000: 387). Television was a new element in their lives, often not seen as relevant, even when the serials were set in rural locations. 'Television is, in this village, one part of a complex jumble of life and the dramatic experiences and visions offered are surprisingly easily incorporated as discrete – not overwhelming – elements in this jumble' (2000: 387).

Unfortunately, much of the research on television and other mass media is largely focused on the issue of globalisation and cultural imperialism and the extent to which global media are dominated by a small number of companies, in particular those based in the US. This discussion has highlighted a number of social and cultural issues that cut across the urban–rural divide. There is a suggestion in some of the empirical research that in the main media tend to be produced in core locations. For example, at a global level, Los Angeles – Hollywood in particular – is a core location for the production of film and television which are broadcast not only in California or the US, but throughout the world. This creates concern for some about cultural imperialism, for others about the dilution of local cultures. Yet others argue that – as in the case of Bollywood –

many of the Hollywood studios are owned by foreign capital anyway. Schilderman (2002) found that newspapers, radio and television were often cited in surveys of low-income urban residents as sources of information, but except for radio, they were rarely ranked highly, because they were widely perceived as providing news and entertainment rather than information that could affect livelihoods. Similar concerns seem to have been expressed across the rural–urban divide, though limited empirical work has been carried out on the impact of urban-based media on rural lifestyles. However, what research has been done suggests that rural people are as likely as any other consumers of media to reframe their 'viewing' or 'reading' within their own cultural context.

The role of reciprocal communications

So-called 'urban myths' are interesting whether or not they are true, for what they say about those who hear them and communicate them to others. The *Courier* reports one such in a recent special issue on ICT:

> A traditional chief in South Africa was asked the following question: 'If you had the possibility of choosing between a telephone line, a school or a clinic for your village, what would you choose?' To which the chief replied: 'the telephone line, so that I could lobby ministers in the capital about the school and the clinic.'
>
> (Anonymous, 2002: 33)

There is a growing awareness that access to information and knowledge is crucial for the advancement of the poor, in particular the rural poor. Much has been written about the opportunities offered by information technology for helping the world's poorest groups overcome the trap of poverty. Indeed the World Bank is increasingly investing in community telecentres that provide communities – many in rural areas – with access to information and communications technology. The suggested benefits of improving communications technology include the provision of information about health (prevention and treatment), environmental management and markets, facilitating the development of democracy, and providing resources for education. For example, the United Nations Food and Agricultural Organisation (FAO) has invested millions of dollars over several years in developing market information systems (MISs) in ministries of agriculture around the world. These MISs collect, record and disseminate market data for strategic agricultural, in particular food, commodities. However, there is a growing realisation that the data have been collected, handled and disseminated in a way that does not benefit the poorer producers (Shepherd, 1997; Poole and Lynch, 2003).

Another example of how communications can help rural areas is by connecting rural social services with urban-based services. Nowhere can this be more beneficial than in the area described as telemedicine. The ITU suggest the following examples of uses for telemedicine:

> - A nurse in the village can consult with a doctor in a city hospital. The nurse can send the doctor a patient's file electronically.
> - A doctor in the village can obtain a second opinion from a colleague elsewhere.
> - A patient can be examined and diagnosed in the village, without being transported to hospital.
> - The patient can be diagnosed more quickly, and treatment can be started.
> - Health-care professionals in remote communities can further their education and knowledge through in-service training and development (distance learning).
> - Mentoring relationships are possible, involving junior professionals in remote communities and experienced senior professionals in urban health centres.
>
> (International Telecommunications Union, 1997: 14)

In addition to telemedicine, ICT can support schools, providing opportunities for children in remote areas to access information for learning and to make contact with students and pupils in other parts of the world. It has opened up opportunities for distance learning not only at school level but, as in the case of telemedicine, for other professional training and continuing professional development of rural-based professionals, such as teachers, lawyers and magistrates. Reciprocity of communication is considered an important aspect of information technologies used for development.

There is certainly a considerable amount of evidence of the paucity of access to telecommunications in developing countries. For example, Lagos state in Nigeria was reported as having 40 per cent of the country's telephone lines in 1994, of which the vast majority were in metropolitan Lagos itself (Abiodun, 1997). Despite the enormous demand in the city and the rest of Nigeria, the number of lines allocated represented 80 per cent of the capacity of 155,000 lines, leaving expansion of telecommunications in the country severely limited. However, in countries across the world the liberalisation of the telecommunications market has attracted inward investment in developments including cellular networks. The result has been a rapid uptake of telephone services as the unsatisfied demand has been absorbed by new operators, which are mostly owned by international telecommunications companies. The developing world markets have shown robust and rapid growth at a time when most telecommunications companies in the developed world have been showing stagnation or declining profitability (ITU, 2001).

Table 5.1 provides evidence of the inequality at a global level in access to ICT. However, governments and special interest groups worry about possible adverse effects of communications technology, for example, providing access to content not previously available. This concern has focused on the availability of pornography on the internet in particular, but it also extends to other issues, such as changes in behaviour and loss of traditional customs, through exposure to other lifestyles.

It is evident that ICT is spreading rapidly in the urban areas of developing countries, such that without deliberate corrective measures the digital divide is more likely to occur between rural and urban areas. However fast new technology spreads, vulnerable and marginalised sectors are likely to remain disadvantaged, particularly where ICT providers such as internet service providers (ISPs) and cellular telecommunications companies chase high-value markets at the expense of areas with low levels of disposable income but

Table 5.1 ***Selected indicators of ICT penetration by country income level***

Country income level	*Telephone main lines per 1,000 people*	*PCs per 1,000 people*	*Internet users per 1,000 people*
Low	26	2	0
Lower-middle	95	10	1
Upper middle	130	24	4
Newly-industrialising	448	115	13
High	546	199	111

Source: World Bank (1999).

perhaps the greatest need. Where ICT is penetrating rural areas, anecdotal evidence of differential access to technology between population strata suggests that a digital information divide may occur even *within* rural communities. ICT will enhance the level of information first among traders, and second among more innovative and entrepreneurial producers. Increasing informational asymmetries (imbalances in access) may result, with adverse consequences for the balance of market power. For example, Lynch (1992) reported on a survey of horticultural farmers in Tanzania which showed that the main source of market information of about two-thirds of respondents was the traders to whom they sold their produce. In this situation, the trader holds the bargaining power.

However, some innovative approaches have been made to spreading appropriate levels of communications technology to more remote areas. For example, companies and NGOs are adapting high technology to meet the needs of the world's poorest, sometimes in international multimillion-dollar initiatives. For example, AfricaOne, a consortium led by the US telecommunications company AT and T, was established to lay high-volume telecommunications cable around the coast of Africa so as to improve the continent's connections to the rest of the world in one well-coordinated intitiative. In this case the large potential volume of use is expected to cover the capital investment. It should also make an increase in intra-regional traffic possible, by reducing the need to route telephone calls and internet hits via Europe, and reducing the existing reliance on more expensive satellite technology. However, the consortium has run into delays and financing difficulties.

An alternative strategy being pursued at the local level is the introduction of small-scale village call centres. This is an approach which has proved successful in a number of countries. For example, approximately 10,000 private telecentres are in operation in Senegal, with an annual turnover of around US$500 million and employing 20,000 (Zongo, 2001). These centres are being used as a model in a range of other countries, for example for the establishment of *teleboutiques* in Morocco, *wartels* in Indonesia, *monocabinas* in Peru and phoneshops in South Africa and Uganda (see Figure 5.2 for an example in a container in South Africa; also Box 5.2). Such telecentres can provide services that range from telephones to e-mail and internet access. With more investment and training, they can also provide communities with the opportunity to construct their own content. The development of locally appropriate content is increasingly being seen as key to the effective use of internet technologies in the developing world, particularly in rural areas. Inappropriate information is provided by urban-based and internationally based services. These centres, called Multi-Purpose Community Telecentres (MCTs), are an approach adopted by the International Development Research Centre's Acacia Project. There is a range of levels, but the key to success is that the telecentres cater for the level of demand at a price that is affordable to the users; that is they are a commercially viable venture.

In Bangladesh the Grameen Bank, famous for its highly successful micro-credit scheme, has introduced a sister company, Grameen Telecom, which uses the same kind of methodology as the small-scale credit bank, focusing on the development of a network of village operators providing access to a mobile phone. In terms of profitability, Grameen's village phone programme brings in three times as much revenue for the telecom company as urban cellular phones (estimated at an average of $100/month versus $30/month) from up to 100 callers per day (Richardson *et al.*, 2000), due to the high number of users for each telephone operator. The ITU's typical rural service model predicts that people will spend not less than 1.5 per cent of GDP per capita on telecommunications services. Bangladesh has identified the expansion of telecommunications to rural areas as a national priority of development. Rural telecommunications can be a significant source of revenue and can also provide an important means for ensuring rural incomes. Richardson *et al.* (2000) found that a significant proportion of callers (42 per cent) registered the purpose of their call as connected to enquiries about remittances from relatives in the urban areas or abroad, especially in Saudi Arabia. A further 44 per cent indicated a social purpose for calls to distant family or friends. Such calls maintain the social and family networks that ensure remittances are sent to the rural households. In addition, 62 per cent of calls are incoming and these tend to be significantly longer than outgoing calls. Thus distant relatives or friends are willing to bear the cost of communication, so access to phones in rural areas does not necessarily place a large additional burden on rural dwellers. Finally, Richardson *et al.* (2000) emphasise that the availability of telephones in rural Bangladesh produces consumer surpluses through savings on alternatives, such as travel to the city. This suggests that there is a potentially significant market in rural areas of even very poor countries such as Bangladesh and that the availability of such facilities may contribute significantly to the rural economy and social welfare of the phone users.

Box 5.2

Models of telecommunications diffusion

The models of diffusion and use of telecommunications in the developing world are resulting in a far more varied application of the technologies available. Many telecommunications providers are seeking ways of delivering telecommunications to as wide a range of the population as possible, and in many cases they are developing into areas that have previously been without any form of such communications infrastructure.

Grameen Phone

The Grameen Bank became internationally famous for its successful micro-credit scheme. It began by loaning rural women $100 for a milking cow. Some of the income from the sale of milk was used to pay off the borrowed money. Grameen Phone works in a similar way: women borrow a Village Phone Provider package costing $350 which is paid off from money made by providing a phone service to the village. The 10,000 phones across the country, serving 500,000 users, earn the village phone operators an average of $2 per day or $700 per year in a country whose GNI per capita is $360 (World Bank, 2003). Grameen Bank became famous for its enormous success in small-scale credit. It has a repayment rate of 96 per cent.

Adapted from http://www.grameenphone.com/.

Vodacom phoneshops in South Africa

The Vodacom container phoneshops (container shops) have a clear business plan – to provide telephone access – and a clear technology model. Most of them are highly visible, being painted in the company's colours (see Figure 5.2). and very successful, and many are financially profitable. They do not claim to be offering a development service and all are privately owned. A phoneshop is a converted container and costs around R24,000 (R11 = £1 at time of writing) to set up. The container is easily installed, as it only needs to be in an area with cellular coverage and to have access to electricity. However, some users of busier phoneshops have complained of long queues and lack of opportunity to expand the container. The owner's investment is an important motivation to successful management, and the entrepreneurial drive is strong. All staff are trained to run the equipment and management systems for the phones. The Vodacom phoneshops are particularly popular due to their low user cost of 60 cents per unit of call time, which means they can undercut many other types of centres. All offer a telephone service, and a few are experimenting with fax and computer facilities, as in Khayelitsha Township, Western Cape. Staffed centres seem to provide a better service and higher reliability than simple payphones.

The phoneshops were established by Vodacom to meet the Community Service Obligations (CSOs) of its licence, and they appear to have been successful. There are currently over 1,800 phoneshops in the country, which are overwhelmingly successful, providing telephone access and jobs. The majority of the phoneshops are in townships rather than rural areas. However, Vodacom believes that it is heavily subsidising the container shops (costs being estimated at 80 cents per unit of call time rather than the 40 cents that Vodacom receives). So, while this is a good model, it is unlikely to develop independently, or without further regulatory pressure.

Another approach is to focus on existing information intermediaries, or *infomediaries,* rather than on the technology (Schilderman, 2002). This approach analyses the existing structures and hierarchies of collecting and organising the information needed for rural economy and society. These may be public service providers, NGOs and community-based organisations, religious groups, some private companies and customary and traditional authority hierarchies. However, these groups have also been accused of 'gatekeeping'

Figure 5.2 ***South African telecom container.***

(restricting access to information to certain groups), promoting their own agenda and circulating inappropriate information (Schilderman, 2002). Then the aim is to support the infomediaries and thus the rural community through effective and appropriate provision of information to meet its needs and demands. The information provided by infomediaries is embedded and verified by the recipients through their social networks. These social linkages may be very important for people on low incomes, because social capital can serve to replace or access financial capital where it is missing. However, Schilderman (2002) suggests that the social networks of people on the lowest income can be very limited. Social exclusion of certain groups can also mean their exclusion from access to accurate and appropriate information. There is evidence of gender differences between men and women, women's social networks often being spatially restricted to the locality, while men's often extend beyond the neighbourhood. This can have implications for verification and accuracy. NGOs, CBOs, state agencies and other infomediaries can be vital to strengthen or supplement limited social networks of information.

A number of initiatives have been developed which are based on the existing skills of poor people and on communicating in language and using media that are easily understandable. One of the ways in which infomediaries have achieved this is through adapting existing forms of cultural expression, such as live performances of music, theatre puppetry or dance. This has been particularly useful where broadcast media are less accessible, or in order to complement messages they provide. This kind of communication relies on the importance of the given type of performance in each culture and the immediacy of word of mouth. It uses a cultural syntax that is familiar to the audiences while communicating knowledge and information that are important to their livelihoods. One of the advantages of this approach is the low level of technology required and therefore the low cost. Another is that more of the target audience can get involved in the communication. For example, it is far easier for marginalised groups to have access to means of communication through culturally appropriate performances than to gain access to national broadcasting facilities that are likely to be indiscriminate and may be culturally inappropriate. 'Theatre for Development' is an association of 26 theatre groups in Zimbabwe which focus their efforts on raising awareness about issues relating to AIDS. The group research the issue with communities, develop a script, and rehearse and then trial performances. Full performances of such pieces are then staged at a range of events and locations including periodic market days, festivals and sporting events.

While international agencies are seeking ways of making ICT available to poor rural and urban dwellers in developing countries, the sale of mobile phones is expanding at very rapid rates (see Figure 5.3), in some cases overtaking landline use. In 1992 Tanzania had 0.29 main phone lines per 100 inhabitants in 1992, compared with 0.77 in Kenya and Uganda at 0.45. Telecommunication infrastructure is concentrated in the cities, leaving remote areas inadequately catered for. In most rural areas of East Africa telephone density (or 'teledensity') is non-existent or quite poor (Mutula, 2001). One consequence of this is that mobile telecommunications companies in developing countries are now seeking highly innovative ways of providing affordable access to their services, experimenting with cellular, radiowave and other technologies.

An example of where information and communication flows between urban and rural areas can have a beneficial effect is in agricultural marketing information. Geographical constraints constitute barriers to information flows, just as much as to physical flows of produce. Therefore rural locations that are remote from urban markets experience informational problems, resulting in

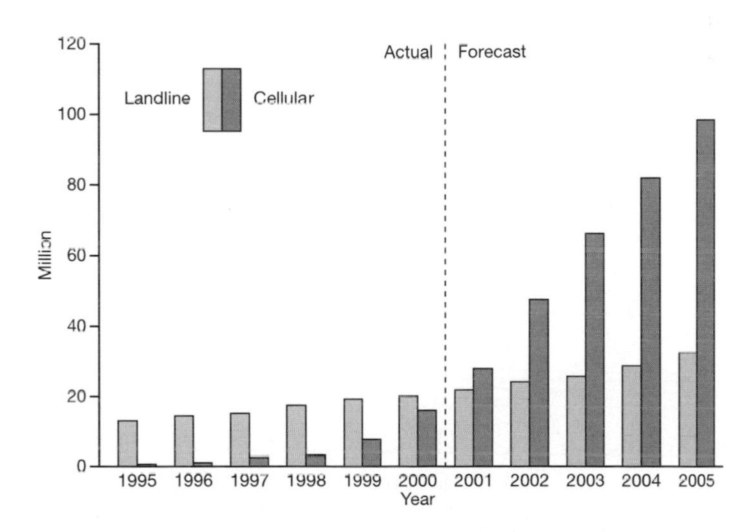

Figure 5.3 ***Graph illustrating the rapid uptake of telecoms.***
Source: adapted from ITU (2001).

un-exploited market opportunities, seasonal gluts and produce with inadequate quality specification and control, inequitable returns to producers, peri-harvest (in-field pre- and post-harvest) losses and fundamentally poor returns to the production and marketing system as a whole. One method of overcoming this problem that is beginning to be applied is the introduction of communications technology. Evidence suggests that improved communication technologies can have an empowering influence on those who have access to them and are able to take advantage of them. In the case of traders, Fafchamps and Minten provide evidence that better-connected traders 'have significantly larger sales and gross margins than less connected traders after controlling for physical and human inputs' (1998: 1). They go on to argue that the social network capital enables traders to deal with each other, mediating relations of credit, trust and flows of information. However, there is also anecdotal evidence that those with access to commercially important marketing information are in a strong position in regard to the market, and that this may be to the disadvantage of those without such access (Lynch, Poole and Ashimogo, 2001).

Figure 5.4 shows Schilderman's knowledge information system model based on empirical research in Peru, Sri Lanka and Zimbabwe. The focus of the model is on the community, represented by the circle in the middle. The eight livelihood criteria are represented by the windmill vanes radiating out from the community. The key informants and infomediaries and their role

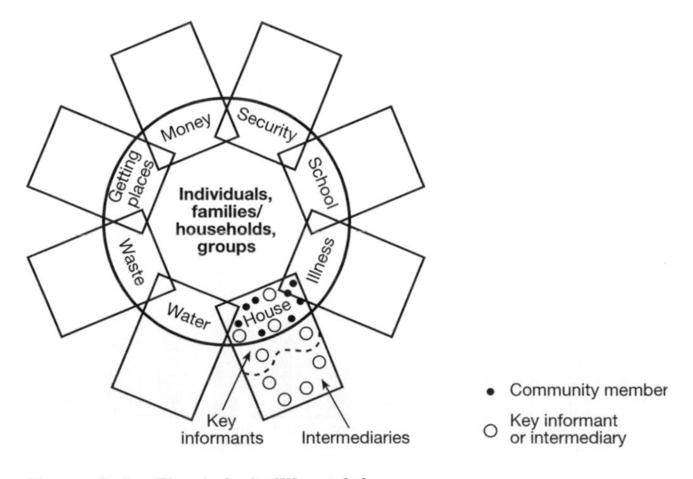

Figure 5.4 ***The 'windmill' model.***

Source: Theo Schilderman, *Strengthening the Knowledge and Information Systems of the Urban Poor*, ITDG, Rugby, March 2002.

in information collection and knowledge generation are depicted both inside and outside the community. In this model, livelihoods are the key focus.

Another model for the diffusion of ICT to rural areas is being promoted by the M. S. Swaminathan Foundation in the state of Madhya Pradesh in India. It is working to develop a network of villages that have access to computers. The state then selects an operator who receives a licence to operate a computer and sell information, which can be accessed from the state's computer network. A charge of $0.25–0.35 is made for printouts of documents such as land records, caste certificates and proof of income, which the villagers would otherwise have to travel to town to obtain. For $0.25, they can email a complaint to the state, which guarantees a reply within a week. Some operators will type out e-mails for villagers, since many are illiterate. Finally, $0.10 is charged for a printout of the prices of any agricultural commodity sold in surrounding markets. Farmers in the villages report that this improves their bargaining position with middlemen. The project was set up in January 2000 and by May that year, the *New York Times* (28 May 2000) reported that 22 villages had joined the scheme.

Such innovative uses of ICT demonstrate the opportunities that are available for making the links between rural citizens and urban-based bureaucracies much more open and transparent. They also illustrate the commercial potential for the poorest rural farmers, who can easily recoup the cost of a 10-cent printout from the better prices they are able to obtain from their improved access to marketing information. However, as these examples illustrate, the innovation that is required is not so much technology as ensuring access to the technology. The examples emphasise an entrepreneurial approach involving a franchise and charging regime, where the operator keeps most of the fee, but pays a proportion to the village and state governments. In this approach the computer operator is given an incentive for maximising use. This is similar to the Grameen Phone scheme (see Box 5.2), where village phone providers offer a call service for a fee most of which they keep, but a proportion of which goes to the phone company to cover call charges and phone lease.

The research by a range of different workers discussed above implies a set of emerging principles, though this is still a relatively new area for research and much of the technology and the models for using it and ensuring access to it are still relatively untested. First, in order for such technologies to have maximum effect in tackling the issue of poverty strategies must be developed which ensure access for those who are economically, socially and geographically marginalised. A second key principle for successful implementation of the diffusion of ICT access involves collaboration between local private and public sectors. Such an approach ensures that private sector initiatives are appropriate to policy frameworks and that public sector initiatives are properly resourced and demand-led. Third, the issue of the projects being demand-led is crucial. The example of market information systems (pp. 143–144 above) illustrated that where initiatives are taken by bureaucrats without consultation with the intended ultimate beneficiaries, the result is that the information provided is often too late, inappropriate or not useful.

This discussion has emphasised the importance of reciprocal communication, and/or ensuring that user groups provide feedback to make it possible for a sharp focus to be maintained on what the users of the information demand and need. Most of the solutions discussed above involve an element of technology and all involve some kind of organisation. In either case, capacity building or training is seen as important to ensure that the technology can be maintained and repaired. It is also important to raise awareness of the potential of the information that is being made available. This will involve public awareness campaigns.

The fifth key principle appears to focus on the technology itself. Although dominant trends in the use of particular technologies are emerging in the developed world, these are not universally applicable. For example, the Grameen Phone demonstrates a use of the mobile phone that is unlikely to occur in affluent societies, where mobile phones, computers and even televisions are assumed to be for individual use only. As this chapter has illustrated, in many developing countries radios, mobile phones, computers and televisions may be shared by communities, with very powerful results. Similarly, communications infrastructure may involve a mix of technologies that are novel to the developed world. For example, mixes of technology, such as cellular, copper-wire, wireless local loop, high frequency radio, microwave and fibre-optic cable, may provide the infrastructure that ensures affordable and effective connection of remote rural areas.

Finally, the examples above have illustrated that women in developing countries, and in particular in rural areas, stand to gain considerably from improved access to information and the media. However, new initiatives have to be aware of the variety of cultural issues involved in access to information. For example, Richardson *et al.* (2000) found that women in rural Bangladesh were more likely to use a phone managed by a female provider than by a male. There is some evidence that women use phones and computer facilities in slightly different ways from men. For example, women use phones to keep in contact with relatives who have migrated to cities or abroad (and thus obtain information about remittances); women may have lower levels of literacy or English (most web-based information is in English); and women may have less power as users to influence content, which is particularly important in broadcast media.

Conclusion

Improvements in access to communications can provide information that empowers, but there is also concern that these improvements may expose populations to exploitative influences. In particular, the issue of communications has been, albeit with limited empirical evidence, linked to existing theoretical debates relating to the trends of globalisation and dependency. Morris (1998a), for example, suggests that globalisation to some extent represents a 'Californianisation' of taste among consumers, emphasising the importance and influence of one region of one country to the global media and communications industry in particular. Some researchers have argued that convergence towards a uniform world view is a concomitant of the development process, and of access to expensive material goods which themselves are standardised.

Discussion questions

- Debate the extent to which the media are important in the fight to eradicate poverty.
- Critically assess the importance of the definitions of 'urban' and 'rural' to the broader issues of development.
- Create a table that compares the advantages and disadvantages of the various media discussed in this chapter.

Suggested reading

Englund, Harri (2002) The village in the city, the city in the village; migrants and Lilongwe. *Journal of Southern African Studies* 28 (1), 137–154.

O'Farrell, C., Norrish, P. and Scott, A. (2000) Information and communication technologies for sustainable livelihoods. *SD Dimensions*. http://www.fao.org/sd/cddirect/cdre0055d.htm [last accessed: 12 March 2004].

Richardson, D., Ramirez, R. and Haq, M. (2000) *Grameen Telecom's Village Phone Programme: a Multi-Media Case Study*. TeleCommons Development Group, Guelph. http://www.telecommons.com/villagephone/index.html [last accessed: 12 March 2004].

6 Finance

Summary

- Financial flows are an important component of rural–urban interaction. There is a diverse range of ways in which finances flow between urban and rural areas and there is evidence that the flows can go in either direction.
- Much research on urban–rural financial flows focuses on the provision of credit to rural areas and the remittances sent by rural-to-urban migrants to their relatives at home.
- Research suggests that both market-based and government-run economies have exploited rural areas for urban gain. This means net wealth flows tend to operate from rural to urban areas. Rural areas have tended to encounter considerable challenges in obtaining access to effective credit and financial services.
- There are examples of successful and sustainable initiatives to provide financial services, including credit, to the low-income residents of rural areas. However, many initiatives suffer from high levels of dependence on subsidy, low levels of loan repayment and poor sustainability. Notable exceptions to this include the Bank Rayat Indonesia (BRI) and the Grameen Bank in Bangladesh.

Introduction

One of the key roles that rural–urban links play in the development process is in the exchange of finance. This can take a number of forms. Finance can be passed to rural areas from urban-based banking and finance institutions in the form of credit provided to farmers and rural businesses. It can also take the form of investment in rural industries from either private sources or government or donor projects. Informal flows of finance may take the form of remittances sent by urban dwellers to their rural-based relatives. One of the themes of this book is that the increased flows between rural and urban areas in the developing world appear to facilitate economic and social activity and can help to build economic and social capital. This chapter focuses on flows of finance and sets out to examine the evidence for their existence, their patterns and the extent to which they can contribute to development in both urban and rural areas.

Economic theories related to rural–urban links

Knight and Song (2000) argue that there are three basic economic theories that apply to economic relationships between rural and urban areas. The first is the Lewis model of economic growth with surplus rural labour. The 'coercive' or 'price-scissors' approach suggests that urban industrialisation draws on abundant rural labour. Capital mobilised from rural productivity gains is invested in industrial development which creates increasing demand for agricultural goods and promotes further improvements in the efficiency of agricultural production. The result, according to the model, is improving dynamic equilibrium in rural–urban terms of trade. This has some resonances with Tiffen's (2003) model discussed in Chapter 1 above (see also Figure 1.2). The second model is of economic growth financed by extracting rural surplus. This is related to the Rostow stages of economic growth model (see discussion in Chapter 1 above and Table 1.6). Thus the growth of the industrial sector is funded by reducing agricultural producer prices, consequently depressing rural income and therefore consumption, and promoting urban industrial capital accumulation. The final model of rural–urban economic exchange is based on the notion that economic policy favours the urban over the rural sector – Lipton's (1977) 'urban bias' thesis. The argument is that urban dwellers exert a considerable influence on government because they are politically aware, more vocal and better organised, and include elites such as bureaucrats. The result is improved urban living conditions, promoting rural–urban migration (see Chapter 4 above). As discussed in Chapter 4, the Todaro model of rural–urban migration is based on the idea that migration will take place if there is a perceived differential between urban and rural incomes. These three theories all suggest capital, or wealth, flows from the countryside to the city as a nation develops. However, some researchers are beginning to find evidence of wealth flows in the opposite direction, suggesting that circumstances may influence the direction of wealth flows between urban and rural areas (see Box 6.1).

An element of dependency theory suggests that wealth flows are rural to urban, while classical economic theory suggests that there is the potential for a 'trickle-down effect' from urban to rural. This was an early focus of the research on interactions between urban and rural areas (Friedmann, 1966; Hirschman, 1958; for a succinct review of these theories and a comparison with Rostow – see Chapter 1 above – see Binns, 2002). The reality is more complex and fluid and depends on local circumstances and changing situations. This is reflected in more recent research (see Box 6.1), which suggests that some households have adopted a strategy of obtaining access to both the income-earning opportunities of urban areas and the productive resources in the rural areas (Potts and Mutambirwa, 1998). However, the impacts are seen in families which have fragmented as a survival strategy, resulting in break-up of the structure of their society due to economic pressures (van Donge, 1992b). The so-called 'multi-locational household' represents an attempt to overcome the difficulties of each location while at the same time retaining its benefits for the whole household. This is similar to Epstein's (1973) concept of 'share families' resulting from his research in South India, where family units lived separately, but agreed to pool resources and expenditure. This suggests that migration in various forms is one of a range of strategies adopted by rural dwellers that may lead to some period of time spent in cities.

> [L]abour migration is indeed a forced livelihood response, although it arises from a complex set of social relations (including relations of debt and dependency) rather than simply ecological crisis and subsistence failure. For others, however, migration proves a positive opportunity to save, accumulate capital or invest in assets.
>
> (Mosse *et al.*, 2002: 60)

De Haan (1999) regards remittances as one of the most important aspects of migration and one likely to provide a 'trickle-down' effect.

Box 6.1

Which way do wealth flows go?

In Zimbabwe during the 1990s, the government enacted an economic structural adjustment policy (ESAP) which was intended to reduce the economic burden of the state and ensure that the economy was able to function efficiently. It is debatable whether Zimbabwe needed such an urgent and draconian approach, as its economic indicators were stronger than those of other sub-Saharan African countries, but there is some suggestion that they were under pressure from the International Monetary Fund. Potts and Mutambirwa (1998) carried out a survey of the impacts of ESAP among recent in-migrants in a high-density residential area of Harare, Zimbabwe's capital. This survey found that the majority of respondents perceived that ESAP had made things worse in both the Communal Areas and the cities. Most respondents (72 per cent) also felt that the worst effects of ESAP were felt in the urban areas rather than in the Communal Areas (CAs) where the majority black rural population live. Potts and Mutambirwa (1998) hinge their explanation on the increase in prices for all goods; urban residents had to buy much more, since rural residents can grow their own food and so avoid grocery bills. The following quote, in response to a question about the biggest problem about living in town, illustrates this:

> Eating bought food. At home we eat what is grown . . . you know a cucumber? [In town] I buy a cucumber for a shilling. A cucumber which is eaten by a baboon at home! Ah no . . . [But here in town] you put a bought thing in your mouth! A chicken's egg is bought for twenty cents! Two bob! An egg! So that's a hard life. You buy firewood. Are trees bought? No. You go and break it in the bush. No . . . they are selling my tree to me. That is what is hard about living in town.
>
> (Barnes and Win, 1992, quoted in Potts and Mutambirwa, 1998: 72)

However, perhaps one of the more interesting results of the survey was that the links between urban and rural areas were highlighted by the respondents, as shown by the following quotes.

> Retrenched people went to the Communal Areas and increased the burden there.
>
> Rural people depend on those working in town.
>
> Retrenchments in town equal retrenchments in the Communal Areas.
>
> Urban workers support the rural poor.
>
> (Potts and Mutambirwa, 1998: 70–71)

These quotes reveal how urban and rural people are affected by rural–urban linkages when the economy is under pressure, in particular when urban residents are retrenched, or made redundant. The results of the research suggest that the rural dwellers were no longer receiving the same level of remittances as previously and were having to absorb extra family members who had been retrenched and would otherwise be living in town. The link to the rural area is characterised by Potts and Mutambirwa (1998: 75) as a 'safety net'.

> Urban destitution is becoming more evident in Harare in the 1990s, particularly for those who have no rural alternative such as the landless, widows and orphans. The safety net of the CAs has always been vital for the urban households and the ESAP era is proving yet again the significance of that safety net.

The research which illustrates the positive benefits of remittances argues that they can fulfil an insurance and an investment role. There is debate about the relative significance of the roles in rural society. One of the difficulties of assessing the relative value of remittances is that small amounts sent from an urban cash-based economy can represent substantial amounts in a cash-starved rural economy.

One of the problems in assessing the economic and financial linkages between urban and rural areas is that the role of the cash economy differs between the two. This is illustrated particularly in Box 6.1, which discusses the rural–urban wealth flows in Zimbabwe during the introduction of economic structural adjustment policies in the 1990s. Satterthwaite (2000) highlights the greater need for cash in the urban economy, identifying some of its main uses:

- for public transport to commute and access essential services. This can be a higher proportion of low-income households' expenditure, especially those located in peripheral settlements;
- for school fees and associated costs such as getting to and from school, buying uniforms, and paying exam fees which tend to be higher in urban areas;
- for housing and associated costs, such as water and other services;
- for access to water – and in some instances to sanitation and garbage collection; these can represent a higher proportion of expenditure among low-income households (see Hardoy *et al.*, 2001);
- for food, which is more expensive for urban households, most of which have to purchase their food needs, in contrast to rural areas where most produce their own;

Box 6.2

Rural–urban income differentials in China

In 2001, 40 children were killed in an explosion as they assembled fireworks in a rural school in order to earn $12 per month. Hundreds of villages in Henan province were infected with HIV and AIDS after receiving infected blood transfusions. Rural child mortality rates are beginning to creep up again.

In early autumn of 2002 the *Financial Times* ran a series of articles on the future for China in which these stories were linked. One of the issues on which this series focused was the rural malaise affecting China while the urban areas experience unprecedented growth. The argument was that these incidents were a dramatic wake-up call telling the Chinese government that it should become aware of the difficulties of the people living in rural areas.

One article presented the case of Guo Hui, a farmer in Yonfeng in the central province of Henan. The farmer faces the natural hazards of soil degradation, drought and locusts, and loss of labour since his son went to work in a factory. Despite these difficulties he and his wife earn Rmb 2,000 (at that time around £155 or US$242) from the sale of the crops they are able to produce. However, the article points out that the state places these high burdens on low-income farmers:

- village upkeep tax;
- public interest tax;
- administrative costs;
- education fee;
- local charities fee;
- militia training fee;
- road repair fee;
- family planning fee;
- agricultural tax;
- water tax;
- education fund-raising tax;
- collection fee – covering the cost of collecting all the other fees and taxes.

Mr Guo says that once he has paid all these fees and taxes he has nothing left. His commercial sales simply go to cover his tax burden. However, he is afraid that the entry of his country into the World Trade Organisation will mean farms like his are exposed to more competition from international trade.

As the cities and their surrounding hinterlands on the coast have benefited from economic growth generated by the manufacturing boom, many people have moved to these areas taking their labour and capital resources with them, in search of employment (see Box 4.4). This has led to the saying 'All of the peacocks have flown to the coast, leaving only the sparrows behind.' The result, according to the *Financial*

Times article, is that farmers suffered an average income decline in each consecutive year from 1998 to 2001.

One of the reasons for this is the increasing tax burden imposed on rural dwellers because the more remote village and town governments have fallen deeper into debt following the failure of their enterprises. Town and village enterprises (TVEs) were encouraged as a means of local development and grew rapidly during the 1980s, but during the 1990s they failed to compete with the dynamic new industries establishing themselves on the coast. Many of these coastal enterprises were able to attract inward investment and developed by selling to large markets nearby. They quickly set up distribution networks in the rural areas where they were able to outperform the smaller TVEs, many of which failed as a result. These failures have adversely affected many of the 40,000 rural credit cooperatives that provide 70 per cent of TVE credit and 80 per cent of peasant loans. Approximately half of the loans are non-performing. Local rural government debt has recently been estimated at Rmb 600 billion, a burden that the rural dwellers have to bear.

According to the article, local government officials are sometimes over-optimistic in their estimates of agricultural productivity, resulting in promotion for them and an increased tax burden for the farmers. This problem has led to conflict, even rioting in some areas. The article concludes that the implications of this 'rural malaise' are potentially very serious indeed. If it continues, then it is likely that more people will attempt to move to the cities in search of employment. The government is not in a strong position to rectify the situation. It is reluctant to take steps which might adversely affect the rapid growth of the coastal cities; it has very little income to invest in the rural areas due to other – mainly urban-based – fiscal difficulties, such as pensions and state banking liabilities, which the article estimates are equivalent to 100 per cent of the country's GDP. Ironically, this is the latest generation of a regime that came to power on the back of a peasant revolt.

Adapted from James Kynge (2002); (see also http://www.ft.com/chinafuture/).

- perhaps for health care which may be more expensive in urban areas; if no public or NGO provision is available, private services have to be purchased. This can be exacerbated by the higher incidence of contagious disease and environmental health problems in urban living conditions;
- for child-care, which may have to be paid for if older members of the household need to earn;
- for payments to community-based organisations, bribes to police, fines when arrested for illegal street vending.

In contrast to its role in urban living, cash tends to be a relatively scarce rural commodity throughout much of the developing world. This is illustrated in Box 6.2 which shows that circumstances in rural

China contrast starkly with the rapid development taking place in Chinese cities. The example of China illustrates how the state can place a double burden on the rural economy. In general terms, the urban-based members of a household may be able to provide relatively small amounts of cash which can still be beneficial to the rural-based household. On the other hand, where necessary, rural-based members of a household can provide the urban-based members with rural resources such as food, woodfuel and so on, which may be in abundance in the countryside and can keep down the costs of urban living.

The rural economy of most developing countries is dominated by small-scale activities. Where trade or production and processing exist, it is mostly small and often informal. 'Formal financial institutions find offering services to this [small enterprise] economy risky, costly and unprofitable' (Rogerson, 2001: 127). According to Rogerson (2001), in surveys of all sectors of economic enterprise, whether urban or rural, access to finance stands out as a key constraint. In Africa most small enterprises have no access to institutional credit and in rural areas this constraint is far more acute. These circumstances have led to the development of micro-credit schemes rather than non-financial services and support which were criticised as expensive and ineffective. Social and kin networks have proved very important as sources of informal flexible loans, gifts, savings, and rotating credit and savings associations. Other informal sources of finance that figure prominently are suppliers and merchants, and such informal credit sources are often used for mobilising household savings and financing small businesses. This is where urban-based households can prove extremely useful to rural cash-poor relatives.

The rest of this chapter discusses research on the different types of financial flows:

- formal, institutional;
- informal, such as migrant remittances;
- investment by government, aid agencies and private finance.

Formal capital flows

Robinson (2001) highlights a debate within the literature on micro-level finance, focusing on the means by which credit is provided to the lowest-income households. These are:

- *the financial systems approach*, which focuses on large-scale outreach to poor, economically active households who will be able to contribute to savings and repay small loans out of household and enterprise incomes. The institutional self-sufficiency of the savings and loan provider is a key principle of this approach, the argument being that this is the only way the huge worldwide demand for micro-credit can be met;
- *the poverty lending approach*, which concentrates on reducing poverty through the provision of credit in combination with a range of additional services, such as education and training, employment, health and nutrition.

Robinson (2001) argues that the huge unmet demand for micro-credit is only likely to be satisfied through the financial systems approach because it does not rely on subsidies.

> Governments and donors cannot finance the hundreds of millions of people who constitute present unmet demand for microcredit services. In addition, the poverty lending approach, as indicated by its name, does not attempt to meet the vast demand among the poor for voluntary savings services.
>
> (Robinson, 2001: 23)

It is through the mobilisation of savings that the credit services can become viable. Under the financial systems approach any subsidies provided by government or donors are used to disseminate lessons and to transfer best practices, while the micro-credit approach uses subsidies to finance their lending (see Box 6.3).

There have been problems with financial institutions that mobilise savings. Hulme *et al.* (1996) identify co-operative rural banks in Sri Lanka as a conduit through which rural savings are channelled towards the capital. 'In effect, the CRBs mobilise rural savings for transfer to Colombo where they have helped to finance the growing budget deficit. They are a major contributor to making finance scarce in rural areas: in 1992 for every Rs100 they took on deposit, only Rs14 was lent out' (1996: 229). Alternatives to the CRBs do exist in Sri Lanka, such as the credit and thrift cooperatives (SANASA) which circulate savings in the form of micro-credit *within* the rural areas and which have a far better record of recovering repayments.

The challenge comes in providing affordable micro-finance. This is the case because the cost of providing small-scale credit can be relatively expensive. However, Robinson (2001) provides evidence

Box 6.3

Micro-credit programmes

Hulme and Mosley (1996a) argue that complex as the process of development is, two ideas about it are broadly accepted:

- capital investment is an important determinant of raising incomes and of economic growth;
- capital markets in developing countries rarely work well.

This presents a major challenge to development initiatives: in the absence of effective private finance, how is capital investment raised for poverty alleviation? In both urban and rural areas the answer has been through the introduction of development finance institutions (DFIs). Hulme and Mosley (1996a, 1996b) carried out a detailed study of DFIs in countries across the developing world. Their findings broadly support their argument. In addition, their work underlines the role of credit for micro-enterprises (including small-scale farming) in raising the incomes of micro-entrepreneurs, reducing poverty and facilitating investment in new technology such as tools or high-yielding seed varieties. For example, the Federation of Thrift and Credit Cooperatives (SANASA) in Sri Lanka is a federation of groups ranging from ten to 700 members who contribute to share capital, attend meetings, save regularly and help to ensure the cooperatives are democratically run. The savings create the basis for lending. Some cooperatives run basic savings and loans services, while others employ staff and effectively establish a local rural bank.

Hulme *et al.* (1996) found in a survey of SANASA borrowers that their mean monthly income rose over the period of their last loan, supporting the conclusion that credit is important to raising incomes. They also found that this credit is creating an increased demand for labour as the borrowers employ more workers to advance the purpose for which the loan was taken. 'As labour is a key to higher outputs and incomes amongst households in the South Asian context, this is an important finding' (Hulme *et al.*, 1996). One of the most important findings concerns the wider social and economic benefits of such institutions. This is an important basis on which to argue for some level of continued support for these finance institutions. For example, reliance on *mudalali* (money lenders) in the nearest towns, who often charge interest rates of 20 to 30 per cent per month, has almost been eradicated. Farmers' reliance on 'in-kind' credit from traders for the purchase of seeds or fertiliser on credit in return for higher prices or lower crop purchase prices has also been reduced. The alternative is market-based bank lending which stays clear of micro-lending and sets conditions that small-scale borrowers such as farmers are unable to meet.

Table 6.1 illustrates the range of credit institutions serving the low-income urban and rural dwellers in the developing world. In the main commercial services such as banks are located in large towns and cities. Urban households that are already circulating higher

Table 6.1 ***Categories of financial and credit services available to lower- and middle-income groups***

Type of financial service	*Categorisation of service*	*Target beneficiaries*
Poverty programmes	Heavily subsidised poverty alleviation	Very poor
Savings facilities for small savers	Subsidised poverty alleviation	Economically active poor and lower middle income
Micro-loans	Semi-commercial credit service	Economically active poor and lower income
Bank loans and savings	Commercial financial services	Lower middle income and above

Source: adapted from Robinson (2001).

levels of finance because of their higher cash incomes and costs are in a stronger position than small-scale farmers for whom the cash economy is important but who are dealing in relatively small amounts of cash. The rural dweller interested in taking a loan therefore has both an economic and a geographical barrier to overcome (see Table 6.1).

Robinson (2001) adopts an alternative approach to the analysis of finance for the poor. She argues first that a wide range of financial services are already available in rural areas and low-income urban neighbourhoods through local traders, employers, landlords, commodity wholesalers, pawnbrokers and various types of private money lenders. However, these various informal money lenders typically charge high interest rates. She argues that poverty is the reason for the introduction of development finance institutions. However, a clear distinction should be made between poverty reduction initiatives and micro-finance initiatives. According to this view, the world is littered with examples of failed DFIs, in large part on account of credit being given to those she calls the 'extremely poor' who are mostly unable to repay it, causing the failure of the credit institutions. Providing subsidised loans (which are usually rationed) to the economically active poor prevents them from having widespread access to commercially available loans and uses scarce government and donor funds that would be better used on other forms of poverty alleviation.

that poor people are willing to pay relatively high rates for micro-loans, mainly because the alternatives are no access to credit at all or borrowing from informal money lenders at much higher rates of interest or with unacceptable in-kind demands. Successful examples of the micro-finance approach include the Indonesian local banking system, or unit *desa*, of the state-owned Bank Rakyat Indonesia, discussed by Robinson (2001), and the BancoSol, analysed by Mosley (1996).

Informal money movements

Informal monetary movements across the rural–urban interface may take a range of forms. They can include rural to rural, rural to urban, urban to rural and urban to urban. Montgomery *et al.* (1996) report that the informal financial markets in Bangladesh were estimated at approximately two-thirds of the total during the 1980s, indicating the importance of this source of credit. By contrast, in Kenya, Buckley (1996) found that only one respondent of his study of rural credit reported using an African private money lender, while the few other cases that were disclosed involved Kenyan-Asian money lenders, with one respondent quoted as saying: 'Asians do it because they were not African' (1996: 285). This suggests that access to rural informal credit sources is highly variable.

The opportunity of initiating informal flows of money from urban migrant to rural relatives is one of the main motivations behind rural-to-urban migration (see Chapter 4 above). In addition, informal migrant remittances are a key impact of migration. Regmi and Tisdell (2002) discuss a number of forms and motivations for sending remittances, arguing that each may have different levels of significance to the rural area:

- regular transfers – where a migrant sends regular sums – often relatively small for the migrant, but with profound benefits in the rural destination; these may be motivated by obligation, an aspiration to inherit land or altruism;
- insurance payments – where urban residents send money to help rural relatives in times of difficulty;
- loan payments – the migrant loans money to the rural relatives, though the expectation of repayment may not be high or may involve not cash but in-kind repayment;
- education repayments – the migrant repays the family for the education he or she received when younger that made the migration possible.

Regmi and Tisdell (2002) found in their analysis of data from Nepal that an average of 4.07 per cent of income was remitted home. However, the range went from 15.56 per cent among those who earned up to 10,000 Nepalese rupees, to 2.25 per cent for migrants with an income of over 50,000 rupees. The absolute amount remitted increased with the income bracket from an average of 889.56 to 2,775.30 Nepalese rupees. They report very different data for other countries, with remittances of 38 per cent for Pakistan and 10–13 per cent for Kenya. More comparative work is required in this area to discern general trends. However, remittances are a very difficult subject to research as considerable trust between researcher and researched is required before details of household finances are divulged.

Other than the amounts and proportion of remittances, researchers have been concerned to find out their economic and social impact. Regmi and Tisdell's (2002) survey found that 55.1 per cent of the remittances were used to supplement household expenditure; in particular, they suggest that this money accounts for the purchase of necessities such as soap, kerosene, clothes, salt, sugar and clothes. As a result of consideration of a number of variables they suggest that the hypothesis – in the Nepalese case at least – that the motivation for sending remittances is to secure inheritance of property, in particular land, may be an important explanatory factor.

Chaudhuri (1993) found evidence that among migrants to Durgapur in India, 86 per cent of remittances sent are to rural areas. Some 44 per cent of the remittances are used for what Chaudhuri describes as consumption spending. This corroborates the findings of Hulme *et al.* (1996a), that much of the credit borrowed from informal sources, such as relatives, money lenders and credit cooperatives, was used for consumption spending, such as investment in dwellings, expenditure for special occasions and bridging the gap between agricultural incomes and production costs. Similarly, Buckley (1996) found that rotating credit schemes in Kenya, or 'merry-go-rounds' as they are known, are typified by consumption investment such as hi-fi equipment. This consumption credit, according to Hulme *et al.* (1996a), has the benefit of protecting household assets and future income that might otherwise have been spent on the purchases. Much of this kind of credit is repaid very quickly. SANASA, the federation of credit and thrift cooperatives in Sri Lanka, has a very high level of loan recovery (Hulme *et al.*, 1996a).

In an in-depth analysis of the relationship between the urban and rural areas of the Yogyakarta Special Region (or Daereh Istimewa Yogyakarta, known as DIY) in Indonesia, Rotgé (2000) studied the development of local community mutual self-help networks. The emergence of these networks, known as *gotong royong*, has meant that ways of sharing and distributing wealth have evolved which provide a very strong incentive for the urban residents to remain closely tied to the rural community. Such ties are an example of what Rotgé describes as centripetal forces which encourage the rural dweller to stay at home. They are countered by centrifugal forces encouraging out-migration, such as dissatisfaction with the local means of earning a living, mainly wet rice cultivation. The result in the villages studied was that the people developed a series of variations of commuting, regular migration and circular migration as a compromise between the centripetal and centrifugal forces. In particular Rotgé and his co-workers found evidence that these relationships involved mutual responsibility between the wealthy and the poor in the communities.

> As members of such a community, individuals need not worry about starvation as long as they remain with their kin. Consequently, it would be a high risk to leave the community for long periods, because of the uncertainty of whether paid employment could support them at their destination.
>
> (Rotgé, 2000: 187–188)

The importance of this strategy became particularly clear when the Indonesian economy collapsed during the 1997 crisis which saw the rupiah fall to 15 per cent of its exchange value against the US dollar, 20 per cent of formal-sector jobs lost and 15 per cent knocked off the value of GDP (Atkinson, 2000). Atkinson (2000) reported that official estimates of those living in poverty had declined to 10 per cent by 1996, a year before the crisis. A year after the crisis, the proportion living below the poverty line was estimated at just under 40 per cent and rising.

Rotgé's (2000) research into villages close to the DIY considered the role of income from remittances sent by circular migrants (see Chapter 4 above for more discussion on circular migration). This research carried out in 1986 found approximately one migrant for every five households in one district near Yogyakarta. Research in this region also found that remittances can raise household income by up to 15.6 per cent. The Gini coefficient (a statistic commonly used to calculate the imbalance of distribution of income) among non-migrant households was 0.39, indicating inequalities of income, but much lower at 0.21, among households benefiting from remittances from a circular migrant. Thus, remittances from migrants appear to reduce the income inequalities that are a characteristic of the rest of the population in the areas studied.

In sub-Saharan Africa, research on livelihood strategies of rural dwellers has found evidence of a virtuous cycle of cross-investment between agriculture and non-agriculture. For example, Madulu (1998) demonstrates that the efforts of farmers in Sukumaland, Tanzania, to acquire ox-carts and bicycles have resulted in increased mobility and productivity in agriculture. This has, in turn, provided an important incentive for agricultural work; he finds evidence that 40 per cent of entrepreneurs received their starting capital from agriculture.

A clear pattern of cross-investment between non-agricultural and agricultural activities was identified in many study sites of a multiple country study led by Deborah Bryceson (1999). In Tanzania's Usambara Mountains, 75 per cent of the respondents invested money earned in agriculture in enterprises, whereas the counterflow was even stronger with 84 per cent deriving some of their agricultural capital from non-agricultural sources (Jambiya, 1998). Similar cross-investment in both directions was found in Nigeria's cocoa-producing zone (Mustapha, 1999). In north-west Nigeria, Iliya (1999) reports that 85 per cent of household heads derived their start-up capital from agriculture or livestock keeping, but only very large households with high incomes invest part of their non-farm revenue in agriculture. Most use these earnings for the purchase of daily needs. The bifurcation of investment strategies between rich and poor households is also evident in the Nigerian savannah village of Nasarawan Doya where 56 per cent of higher-income households and 16 per cent of middle-income households derived start-up capital for non-farm activities from agriculture (Meagher 1999). In Nigeria's Middle Belt there was relatively low cross-investment, as only 10 per cent of the non-farm entrepreneurs reported using their earnings to fund agriculture. The agricultural investment cycle here is therefore largely internal to the farming process (Yunusa, 1999). Nonetheless, Yunusa stresses that the rural non-farm sector vitally depends on the purchasing power generated by agricultural proceeds and there was evidence that during periods of economic pressure the rural-based non-farm activities were used by urban-trained non-farm operators as a means to mobilise rural capital, after which they may move the business to the city (Yunusa, 1999; Meagher, 1999).

In Latin America, Tacoli (1999) found that Paraguayan farmers on the urban fringe of Asunción were not able to benefit from proximity to the urban financial markets. This was because their plots of land were too small to be used as collateral for a loan to invest in their

Box 6.4

Cocoa trade and finance flows in Ghana

Ghana in West Africa is one of the world's leading cocoa producers. It produces over 400,000 tons of high-quality cocoa per year. In 1999 cocoa was estimated to be generating 13 to 14 per cent of the country's GDP, 11 per cent of its tax revenue and 30 to 35 per cent of its foreign exchange earnings. It accounts for approximately half of the economically active population and is grown in six of the country's ten regions. Cocoa is therefore of enormous economic and political significance to the people of Ghana. But how are the contributions and benefits of cocoa production distributed between urban and rural areas?

In Ghana, one of the main reasons why rural employment in the cocoa sector is so extensive is that most producers are small-scale farmers who also sun-dry and ferment their cocoa harvest themselves. This processing gives Ghana's cocoa a better taste and higher butter content than the plantation-produced and mechanically dried cocoa of countries such as neighbouring Côte d'Ivoire. The higher quality can bring Ghana $70 per tonne more than other countries in the region. Thus in Ghana more people are employed in production and they receive a better price than in neighbouring Côte d'Ivoire. The floating of the cocoa price in Côte d'Ivoire when the world price fell has been linked to the political unrest in that country in recent years.

With a backdrop of an insecure banking market and limited affordable credit available – in some cases more than 40 per cent interest – to Ghanaian small-holding rural dwellers the income from cocoa production provides an important inflow of cash to the country's rural households from the international markets. The high cost of credit hampered modernisation in the agricultural sector and is partly responsible for the low levels of fertiliser and other inputs used. This is particularly the case since subsidies on inputs were removed under pressure from the World Bank. However, the cost of credit also means that these producers are more dependent on the decisions taken by the government about the prices to be paid to producers. The result is that although the farmers are able to avoid urban-based banks that charge high rates for credit, they are dependent on the decisions made by government in the capital about the prices to be paid. The cities are therefore vital to the livelihood of the farmers.

Ghana came under pressure from the international trade and financial institutions in the late 1990s to reduce the level of state intervention in its cocoa market. Despite the licensing of 18 private traders and the prospect that up to 30 per cent of cocoa would be exported privately, the main issue of concern to the international institutions was that the state cocoa marketing board, Cocobod, was paying farmers more than the planned price, thus a higher price than farmers were being paid in neighbouring countries. The price paid was higher than could be sustained by the higher premium attached to high-quality Ghanaian cocoa. For example, in 1999, despite oversupply and slow demand, Cocobod paid farmers the same price as it had in 1998. The experience of privatisation of cocoa marketing in nearby countries such as Nigeria and Cameroon is instructive in that a drop in production and quality followed. Farmers' representatives argue that guaranteed prices for this international commodity are vital to ensure that national production increases in line with the target of 500,000 tons by 2004. This policy of supporting national production has also provided an incentive to rehabilitate flagging cocoa production and has attracted farmers back to this activity. In particular, it is vital if younger farmers are to be attracted to the sector – the average age of Ghana's farmers is 55 – and to provide an economic base for Ghana's rural population.

The question is: should governments be involved in supporting the production of an international agricultural commodity like cocoa? The indications are that the country benefits: notably, farmers have some assurances about prices and a market outlet, and this encourages younger farmers to remain in the rural areas and grow cocoa, rather than migrate to the cities in search of employment. It is clear that Ghanaian cocoa farmers rely on access to finances from institutions located in the cities, whether these are government or private. By retaining their role in processing the cocoa themselves, the Ghanaian farmers create greater wealth for themselves, bring employment to the rural areas and earn foreign exchange for their country. This inflow of money is vital at a time when Ghana is still very dependent on the export of this commodity, whose price trend is one of long-term decline.

Adapted from Wallis (1999a, 1999b, 2000).

business and take advantage of the proximity to the urban markets for food. She reports that in many cases they are forced to sell to larger farmers. 'The optimal use of productive and physical capitals is therefore intrinsically linked to access to financial assets' (1999: 8). However, she goes on to enumerate social resources (networks of friends, relatives and business associates) and cultural capital (perceptions of creditworthiness, reliability and trust) that could be mobilised to secure credit. In Bangladesh, Afsar (1999) found that improvements in communications and transport have led to an increased incidence of what she calls 'reverse investment' as urban migrants return to provide capital and take part in decision making about agricultural activities. In Dhaka, more than a third of slum residents had cultivable land and around a quarter derive regular income from it. There has also been an increase in rural manufacturing and other non-farm economic activities, with growth rates of 10 per cent for rural processing and manufacturing and 5.5 per cent for rural services between 1988 and 1996. This forms one of the strands of Afsar's (1999) argument that contrary to orthodox migration theory, rural–urban migration does not always adversely affect the rural areas. However, she points out that the communications and transport infrastructure is crucial to ensuring that the benefits return to the rural areas.

Bryceson's (1999) 'betwixt and between' conceptualisation of the rural household emphasises the fact that no two households earn their livelihood in the same way (see also Bryceson, 2000, for an accessible summary of this work). Thus policy interventions that emphasise a particular livelihood strategy are unlikely to benefit all. This can be considered as analogous to the livelihood strategies that straddle the rural–urban interface. The empirical evidence presented by Bryceson (1999) suggests that the mixing of agricultural and non-farm employment can result in flows of investment between these activities. Similar outcomes can result from flows of finance between rural and urban areas. In addition to these arguments, de-peasantisation may also be an explanation for the development of non-agricultural activities in African rural areas that sheds light on the relationship between cities and the countryside at a more general level. This research is mirrored by similar findings in the South-East Asian context (Afsar, 1999; Rigg, 1998a, 1998b) and in Latin America (Reardon *et al.*, 2001; Tacoli, 1999). The bulk of the interpretation explains the development of non-farm employment in terms of:

- declining employment and earning opportunities in cities under structural adjustment programmes;
- the consequent presence of returned migrants in rural areas;
- the low esteem in which younger people in rural areas hold agriculture as a means of income earning;
- the decline of rural services as a result of structural adjustment;
- the low viability of using agricultural inputs to boost productivity;
- growing population pressure.

Government investment and aid

There are numerous examples of ways in which governments have used income from agricultural production to swell the size of their bureaucracies. Box 6.4 illustrates the importance of cocoa to the economy of Ghana. However, the use of rural surplus to support an urban-based bureaucracy is also illustrated by the example of Ghana. Toye (1991) reported that in 1985 Jerry Rawlings, president of

Ghana, condemned the urban exploitation of rural cocoa producers. This condemnation was followed shortly afterwards by a reduction of 16,000 in the staff of Cocobod, the national cocoa marketing organisation; then 10,000 'ghost workers' were removed from the payroll by early 1987 and a further 14,000 were 'retrenched', that is made redundant, by the end of 1987. While this was part of a wider programme of government retrenchment, the implication in Toye's (1991) report is that it was intended to reduce the urban-based burden placed on rural cocoa producers. Despite these retrenchments, at the time Toye was writing in the early 1990s, some observers still thought that the remaining 50,000 Cocobod employees constituted too heavy a burden for the cocoa-producing sector.

Many studies have demonstrated that research in agriculture can yield economic returns, providing improvements in production, post-harvest handling and other areas of agricultural business. For example, the links between the rural and urban economies are illustrated by the statistic that every $1 of additional investment in African agriculture leads to an extra $1.50 of non-farm output while in Asia the figure is $1.80 (Rakodi, 2002). Many such studies on the developing world highlight the positive impacts of agricultural research on rural poverty. Moreover, Fang *et al.* (2003) find that in China there is a positive relationship between government investment in agricultural research and reducing urban poverty. One of the main causal links in this relationship is that Chinese urban dwellers living on less than $2 per day spend 58 per cent of their income on food. Fang *et al.* conclude that in 1992, for every 1 per cent decline in the cost of food, urban poverty is reduced by 3.08 per cent. As might be expected, investment in agricultural research results in a reduction in poverty in rural areas, but their data suggest that for every 10,000 *yuan* of agricultural research investment 10.94 people are raised above the $2 per day poverty line and that by 1998 this amount of investment was raising 6.31 people above the poverty line. This means a total of 4.68 million urban residents were raised above the poverty line in 1992 and 3.16 million in 1998. They argue that the decline in impact on poverty reduction between 1992 and 1998 is mainly due to the declining proportion of income spent on food as it becomes cheaper and more widely available (Fang *et al.*, 2003). However, they argue that continued agricultural investment is required in order to ensure that both urban and rural areas continue to reduce levels of poverty: 'with rapid urbanization, agricultural research will still need to play a key role in supplying adequate food at affordable prices to ensure that urban and rural poverty remain low' (Fang *et al.*, 2003: 740). This begs the question of whether investment in urban and industrial development has the same multiplier effect in rural areas. As discussed above, Afsar (1999) suggests that at least in the case of Bangladesh the benefits of urban income can be passed on to rural areas where social and familial links are maintained by rural-to-urban migrants and where these links are facilitated by good communications and transport infrastructure. Similar findings are reported by other authors such as Potts (1997) in sub-Saharan Africa, Rigg (1998a, 1998b) in Thailand and South-East Asia generally, and Rotgé (2000) in Indonesia. In the latter case comparative village surveys found that proximity to a main road increased the economic and social benefits of linkages with the urban area.

Rutherford *et al.* (2002) found in a survey of the use of financial services over a full year by urban residents in Bangladesh that none had used less than five separate types of service during that period. One had used as many as 18 and the average was ten (2002: 115). Attwood and Baviskar (2002) have two key concerns about the results of research on the way development finance institutions (DFIs) operate in India. The first is that countries like India have poured considerable amounts of capital into DFIs, whose assets rarely consist of members' savings. The credit is mainly provided to landowners, who are mostly men, and the level of repayment is low. 'Thus, formal "co-operative credit" is neither self-supporting nor beneficial to the poorest villagers, including women' (Attwood and Baviskar, 2002: 168). The second key concern is that many of the DFIs overlook the need for services other than credit. Services such as micro-savings and insurance are in demand among the poor and represent an important means by which they can reduce their vulnerability to adverse events.

Formal financial institutions have a role to play in facilitating the flows of capital between urban and rural areas and between farm and non-farm sectors. However, the literature is replete with examples of them failing to achieve this, on account of corruption, incompetence or poor structure. Moreover, there are many cases where poor rural savers and producers have their surpluses expropriated for the development of industrial and urban development. Careful empirical studies, however, appear to show notable examples of successful approaches that focus on the needs of the user, on ensuring long-term financial viability and not restricting activities to the provision of credit, but providing savings, insurance and other financial services at an appropriate scale for the rural residents.

Conclusion

Theories of the nature of financial flows between urban and rural areas tend to focus on ideas of urban extraction of rural wealth and economic assets. Past research has provided evidence of exploitation of rural areas by both urban private interests and urban-biased governments. However, governments and aid donors have attempted to balance this bias by providing rural credit. Although the barriers to obtaining credit in rural areas are considerable, there are some examples of successful rural financial services. Evidence suggests, however, that these are a minority; most rural credit institutions, while providing small but welcome capital to cash-starved rural economies, have a poor track record. There is potential to develop a wider range of financial services including not only credit but also savings and insurance services. However, it is important to the success of such schemes that control remains in the rural area with the clients in order to avoid transfers in favour of the already more solvent urban economies.

Discussion questions

- Critically assess the factors for success in rural credit schemes.
- Review the main ways in which capital has been extracted from rural areas.
- Critically assess the role the state can play in facilitating the movement of finances between urban and rural areas in a way that is beneficial to both.

Suggested reading

Bryceson, D. F. (2000) *Rural Africa at the Crossroads: Livelihood Practices and Policies*. Natural Resources Perspective No. 52. Overseas Development Institute, London. http://www.odi.org.uk/nrp/52.html [last accessed: 14 March 2004].

Murray, C. (2002) Rural livelihoods. In V. Desai and R. Potter (eds) *Arnold Companion to Development Studies*. Edward Arnold, London. 151–155.

Reardon, T., Berdegué, J. and Escobar, G. (2001) Rural nonfarm employment and incomes in Latin America: overview and policy implications. *World Development* 29 (3), 395 – 409.

Rigg, J. (1998a) Tracking the poor: the making of wealth and poverty in Thailand (1982–1994). *Journal of Social Economics* 25 (6/7/8), 1128–1141.

Robinson, M. (2001) *The Microfinance Revolution; Sustainable Finances for the Poor*. World Bank, Washington, and the Open Society Institute, New York.

7 Conclusion and future perspectives

The other chapters in this book have concentrated on a range of aspects of specific relations between cities and the countryside in the developing world. Throughout the book an attempt has been made to examine the nature of rural–urban relationships within the broader context of the role of these relationships in general ideas of development. One of the arguments in this book is that rural–urban interactions have been an important part of the general perception of development in the past and that they have been coming back into focus again in recent years. Researchers, planners and development organisations are establishing initiatives and activities that attempt to use the dynamism of the interactions for the benefit of both urban and rural areas. However, one of the main challenges to successfully harnessing urban–rural linkages for the purpose of mutual or interdependent development is the fact that there is still much to be learned about the nature of these relationships and there are competing views on how best this can be done.

Conceptualising rural–urban dynamics

Chapter 1 highlighted the range and complexity of the relationships between cities and countryside. This book has attempted to clarify some of the key ideas in order to be able to develop a better understanding of them. One general aspect of much of the research that has been discussed in this book is that, in the main, rural–urban linkages have been secondary. The main focus of discussions of credit provision and food production has been on how the linkages can benefit low-income rural producers and dwellers. Discussions on food supply, the peri-urban interface, energy systems and water have largely been focused on the systems supplying urban areas. The main exception to these bodies of work which focus on either rural or urban is research on migration, which places the greatest emphasis on rural–urban linkages, in that the explanation of patterns of migration has to branch out into considering flows of wealth and ideas of 'urbanism' and 'rurality', how they are disseminated and so on. Rider Smith (1999) suggests that one of the reasons for the dichotomy between rural and urban is the difficulty of finding a 'home' for initiatives linking the two. For example, many national, international and non-governmental organisations have departments or sections that focus on either urban or rural issues. There is therefore now a pressing need for a better understanding of the relationships between cities and rural areas within a context that is bound not by a focus on one or the other, but on the linkages between them. A growing emphasis on the livelihood strategies of the people who themselves often straddle the divide provides a useful starting point for a more integrated approach to this theme. For example, Satterthwaite (2000) conceptualises the rural–urban continuum as shown in Table 7.1.

This book has tried to cover a considerable amount of detail in a relatively compact space. One danger of writing a book on a set of topics that affect a very large proportion of the world's population is that either the writer is guilty of over-simplifying or the issues are so big that the reader finds it difficult to understand them without trying to simplify what he or she has read. One of the most important conclusions therefore should be that the issue of the linkages between city and rural hinterland in the developing world is made up of a highly complex set of relationships which vary considerably from country to country and region to region. This said, because of the fact that there are a number of common themes that affect the relationships, there are some generalisations which can be cautiously made. Most cities in the developing world are undergoing very rapid growth. In many countries both cities and rural areas underwent a period of colonisation which established an infrastructure to facilitate the flows of goods, people and capital, in a way that favoured the exploitation of the rural by the urban and ultimately of both by the colonising power. Many countries have tried to overcome this bias through reorganisation of their infrastructure and their society or by

Table 7.1 ***The rural–urban continuum***

RURAL	RURAL–URBAN INTERFACE	*URBAN*
Livelihoods drawn from crop cultivation, livestock, forestry or fishing (i.e. key for livelihood is access to natural capital)		**Livelihoods** drawn from labour markets within non-agricultural production or making/selling goods or services
Access to land for housing and building materials not generally a problem		**Access to land for housing** very difficult; housing and land markets highly commercialised
More distant from government as regulator and provider of services		**More vulnerable to 'bad' governance**
Access to infrastructure and services limited (largely because of distance, low density and limited capacity to pay?)		**Access to infrastructure and services** difficult for low income groups because of high prices, illegal nature of their homes (for many) and poor governance
Less opportunities for earning cash; more for self provisioning; greater reliance on favourable weather conditions		**Greater reliance on cash** for access to food, water, sanitation, employment, garbage disposal, etc.
Access to natural capital as the key asset and basis for livelihood		**Greater reliance on house as an economic resource** (space for production, access to income-earning opportunities; asset and income-earner for owners - including *de facto* owners)
Urban characteristics in rural locations (e.g. prosperous tourist areas, mining areas, areas with high-value crops and many local multiplier links, rural areas with diverse non-agricultural production and strong links to cities)		Rural characteristics in urban location (urban agriculture, 'village' enclaves, access to land for housing through non-monetary traditional forms, etc.)

Source: Satterthwaite (2000).

attempting to develop intermediary settlements and thus shift the focus of development away from the core. These strategies have had varying degrees of success.

In addition to the above context, we can also propose conclusions about the future of rural–urban relations. In the short to medium term, it seems that the process of urbanisation is inevitable, because of the pressures that will be brought to bear on the rural areas of the developing world and because of the greater opportunities available in the developing world's cities. The evidence suggests that over the next 25 years the world's population will become predominantly urban based. Much of the transition will take place in the developing world, where people will move into cities because rural living is no longer sustainable or because the urban environment provides better opportunities.

However, while urbanisation is inevitable and will be a particular feature of the developing world, there is also the process of counter-urbanisation. For the moment this is a process that is only apparently widespread in the advanced economies. However, there is evidence of people moving out of cities in processes that have been termed 'return migration' whereby people who moved into the city for work return after a period of time. 'Circular migration' is the pattern in which rural dwellers migrate to a city or to a number of cities, perhaps working their way up the urban hierarchy, but with the ultimate aim of earning enough income to return to work on their farm back in their home village.

Policy to enable rural–urban interaction reduces poverty

Until now it has been rare for any policy maker to consider the role of urban–rural interaction as an area that could provide opportunities for intervention with a view to poverty reduction. Most policy has focused on either urban or rural areas, but rarely on both. However, this book has set out to demonstrate that there is now a considerable and growing body of evidence to support the notion that rural–urban relations result in positive development outcomes. Unless carefully managed, these can also increase the vulnerability of the urban and rural poor. These relations therefore merit serious consideration for the likely impacts of future policies focused on either urban or rural areas but also for the possibility that positive policy development on these themes has the potential to provide considerable development benefits. However, another key conclusion drawn from this review of the research is that the nature of urban–rural linkages varies from one region to another and from one country to another. This means that it is almost impossible to recommend policies that are appropriate to all circumstances. What most research suggests, however, is that future policy development should be formulated taking into account the possible impacts and likely opportunities of rural–urban interaction. This is likely to require open consultation with the people affected, using a process that focuses on the livelihood strategies of those whose economic and social activities straddle the rural–urban continuum.

It is instructive that as development initiatives in the world's poorest countries explore the opportunities that rural–urban interactions can provide, similar exploration appears to be under way in the most advanced economies. Analysis of the changes taking place in Los Angeles by writers such as Soja (1989) and Davis (1990) contains the suggestion that as urbanisation proceeds in the world's poorest countries, such urban dominance is beginning to relax among the world's richest. The process of 'counter-urbanisation', or the movement of people out of urban centres, is one of several trends (Pacione, 2001) that are increasingly evident in cities, such as the development of 'edge cities', 'suburbanisation' and, with the development of telecommunications, 'the informational city' (Castells, 1989). 'Edge cities' are a form of urban development that involves refocusing existing cities around shopping malls that are usually on green-field sites on the outskirts of cities. One implication of this transition is that the central places of public circulation are essentially controlled and privatised. In some situations the management of the edge cities is even privatised and run by town-centre management companies. These new types of cities grow up as a result of disillusionment among the middle class or college-educated about the role of the city as economic powerhouse. Similarly, suburbanisation represents the expansion of urban lifestyles into rural areas in an attempt to recapture the rural idyll as a backdrop to urban mass-consumption lifestyles. The concept of the 'informational city' envisages that cities remain key locations of the decision making of the global economy, but the power of information technology renders space and place meaningless outside networks of social, technological, cultural and economic activities. The result is that advanced cities are increasingly deindustrialising and restructuring around increasingly globalised networks where information is the strategic commodity. This argument suggests that historical differences between cities are increasingly being internalised by cities and their appendages, edge cities and suburbs. The international flows of people, goods, resources and capital become ever more important and the world is divided into the 'information rich' and the 'information poor', who are increasingly present in all cities, though in differing proportions. These new ideas about different types of built landscape are attempts to organise thinking around the changes that are taking place in rural and urban areas and in the relationships between them.

Themes of rural–urban relations are a vitally important focus for future research and policy development. In order to develop sustainable cities and sustainable rural areas in the future, it will be necessary to strengthen the linkages between the cities and their rural hinterlands. The proof of this is found partly in the arguments of the many international agencies that are looking to the future to see how the rapidly growing cities of the developing world can be sustainable. The proof is also partly found in the fact that households in developing countries are already using their own intra-household rural–urban relations to maximise their livelihood strategies. Establishing links between the city and the country is therefore seen as a strength and the reason for this is the complementary needs and opportunities of the town and country that have spanned millennia. Governments can learn from these households. The challenge will be to enable the populations of the developing world to manoeuvre themselves into that position of strength.

In summary, there is a number of interventions in which government and agencies can engage. These could include the strengthening of rural–urban relations through facilitating marketing, transport and communications, the provision of infrastructure for small and medium towns as well as large conurbations, encouraging private and public services to serve the rural as well as the urban population, and strengthening local and regional governments at all levels. However, it is important that these are carried out together. It is important to be aware that there is evidence that improving communications and transportation infrastructure, for example, is as likely to result in the exploitation of a rural area as in its development. Beneficial development intervention, therefore, depends on a better socio-economic understanding of the likely benefits and on some kind of stakeholder representation in the design and implementation of any interventions. This will ensure that 'blueprint solutions' are not applied inappropriately to different regions.

The evidence therefore suggests that rural–urban relations are very potent. Like all relations, they can be mutually beneficial at one extreme or entirely exploitative at the other. There is tremendous potential for strengthened rural–urban relations to result in development and poverty reduction. However, this has to occur in a context where those involved have the social, economic and physical resources to take advantage of such opportunities to ensure their sustainable urban and rural livelihood.

References

Abiodun, Josephine Olu (1997) The challenge of growth and development in metropolitan Lagos. In C. Rakodi (ed.) *The Urban Challenge in Africa: Growth and Management of Its Large Cities*. United Nations University Press, Tokyo. 192–222.

Abu-Lughod, Lila (2000) The objects of soap opera: Egyptian television and the cultural politics of modernity. In Kelly Askew and Richard Wilk (eds) *The Anthropology of Media: a Reader*. Blackwells, Oxford. 376–393. (Reprinted from *Public Culture* 5 (3), 1993, 493–513.)

Afsar, R. (1999) Rural–urban dichotomy and convergence: emerging realities in Bangladesh. *Environment and Urbanization*. 11 (1), 235–246.

Allen, A. (2001) Environmental planning and management of the peri-urban interface. Keynote paper presented to the conference: *Rural–urban Encounters: Managing the Environment of the Peri-urban Interface*. Development Planning Unit, University College London, 9–10 November 2001.

Amselle, J.-L. (1971) Parenté et commerce chez les Kooroko. In C. Meillassoux (ed). *The Development of Trade and Markets in West Africa*. Oxford University Press for the International Africa Institute, London. 253–265.

Anonymous (2002) Information and communication technologies (ICTs). *The Courier* 192 (May–June), 33. http://europa.eu.int/comm/development/body/publications/courier/index_192_en.htm [Last accessed: 12 March 2004]

Aragrande, M. (1997) *Methodological Approaches to Analysis of Food Supply and Distribution Systems*. Food into Cities Collection, AC/11–97E. Food and Agriculture Organisation, Rome. http://www.fao.org/ag/ags/agsm/SADA/DOCS/DOC/AC1197E.doc [last accessed: 30 November 2003]

Atkinson, Adrian (2000) *Promoting Sustainable Human Development in Cities of the South: a Southeast Asian Perspective*. Occasional Paper No. 6. United Nations Research Institute for Social Development (UNRISD), Geneva.

Attwood, D. W. and Baviskar, B. S. (2002) Rural co-operatives. In V. Desai and R. Potter (eds) *The Companion to Development Studies.* Arnold, London. 165–169.

Bakker K. (2003) Archipelagos and networks: urbanization and water privatization in the South. *The Geographical Journal*, 169 (4), 328–341.

Barnes, T. and Win, E. (1992) *To Live a Better Life: an Oral History of Women in the City of Harare, 1930–70.* Baobab Books, Harare.

Barrett, Hazel R. (1986) The evolution of the marketing network in the Gambia, in the twentieth century. *Tijdschrift voor Economische en Socialische Geografie* 77 (3), 205–212.

Beall, J. (2002) Living in the present, investing in the future – household security among the poor. In C. Rakodi with T. Lloyd-Jones (eds) *Urban Livelihoods: a People-Centred Approach to Reducing Poverty.* Earthscan, London. 71–87.

Binns, T. (2002) Dualistic and unilinear concepts of development. In V. Desai and R. Potter (eds) *The Companion to Development Studies.* Arnold, London. 75–80.

Binns, T. and Lynch, K. D. (1998) Sustainable food production in sub-Saharan Africa: the significance of urban and peri-urban agriculture. *Journal of International Development* 10 (6), 777–793.

BMA (2001) *Bangkok State of the Environment 2001.* Bangkok Metropolitan Authority (BMA), Bangkok. Available at United Nations Environment Programme website http://www.rrcap.unep.org/reports/soe/bangkoksoe.cfm [last accessed: 1 December 2003].

Bohannan, P. and Dalton, G. (eds) (1962) *Markets in Africa.* Northwestern University Press, Evanston, Ill.

Boserup, E. (1965) *The Conditions of Agricultural Growth.* George Allen and Unwin, London.

Bradford, A., Brook, R. and Hunshal, C. S. (2003) Wastewater irrigation in Hubli-Dharwad, India: implications for health and livelihoods. *Environment and Urbanization* 15 (2), 157–170.

Briggs, J. (1993) Comment: population change in Tanzania – a cautionary note for the city of Dar es Salaam. *Scottish Geographical Magazine* 109 (2), 117–118.

Briggs, J. and Mwamfupe, D. (2000) Peri-urban development in an era of structural adjustment in Africa: the city of Dar es Salaam, Tanzania. *Urban Studies* 37 (4), 797–810.

Bromley, R. D. F. (1998) Market-place trading and the transformation of retail space in the expanding Latin American city. *Urban Studies* 35 (8), 1311–1334.

Brook, R. and Dávila, J. (2000) *The Peri-Urban Interface: a Tale of Two Cities.* School of Agricultural and Forest Sciences, University of Wales, and Development Planning Unit, University of London.

Brown, L. (1997) *Agriculture – the Missing Link.* Worldwatch Paper 136. The Worldwatch Institute, Washington, DC.

Bryant C. R. and Johnston, T. R. R. (1992) *Agriculture in the City's Countryside*. Belhaven Press, London.

Bryceson, D. F. (1985) Food and urban purchasing power: the case of Dar es Salaam, Tanzania. *African Affairs* 84 (337), 499–522.

Bryceson, D. F. (1987) A century of food supply in Dar es Salaam. In J. I. Guyer (ed.) *Feeding African Cities*. Manchester University Press, Manchester. 154–202.

Bryceson, D. F. (1993) *Liberalizing Tanzania's Food Trade: Public and Private Faces of Urban Marketing Policy 1939–1988*. UNRISD with James Currey, London.

Bryceson, D. (1999) *Sub-Saharan Africa Betwixt and Between: Rural Livelihood Practices and Policies*. De-Agrarianisation and Rural Employment (DARE) Network. African Studies Centre Working Paper 43. University of Leiden.

Bryceson, D. F. (2000) *Rural Africa at the Crossroads: Livelihood Practices and Policies*. Natural Resources Perspective No. 52. Overseas Development Institute, London. http://www.odi.org.uk/nrp/52.html [last accessed: 14 March 2004].

Bryceson, D. F. (2002) The scramble in Africa: reorienting rural livelihoods. *World Development* 30 (5), 725–739.

Buckley, G. (1996) Financing the Jua Kali sector in Kenya: the KREP Juhudi scheme and Kenya Industrial Estates Informal Sector Programme. In D. Hulme and P. Mosley (eds) *Finance against Poverty*, Volume 2. Routledge, London. 271–332.

Burke, J. and Beltran, J. (2000) Competing for water. *Habitat Debate* 6 (3). Available at http://www.unhabitat.org/HD/hdv6n3/competingforwater.htm [last accessed: 13 July 2004].

Castells, Manuel (1989) *The Informational City: Information Technology, Economic Restructuring and the Urban–Regional Process*. Basil Blackwell, Oxford.

Chambers, Robert (1982) *Rural Development: Putting the Last First*. Intermediate Technology Development Group, London.

Chambers, Robert (1997) *Whose Reality Counts? Putting the First Last*. Intermediate Technology Development Group, London.

Chan, K. W. and Zhang, L. (1999) The Hukou system and rural–urban migration in China: processes and changes. *China Quarterly* 160, 818–855.

Chant, S. (1998) Households gender and rural–urban migration: reflections in linkages and policy considerations. *Environment and Urbanization* 10 (1), 5–12.

Chaudhuri, J. R. (1993) *Migration and Remittances: Inter-Urban and Rural–Urban Linkages*. Sage, New Delhi.

Clark, V., Baker, J. and Lewis, E. (2002) *Key Concepts and Skills for Media Studies*. Hodder & Stoughton, London.

Club du Sahel (1994) *West Africa Long Term Perspective Survey*. Club du Sahel, Organisation for Economic Cooperation and Development, Paris.

Cohen, A. (1971) Cultural strategies in the organisation of trading diasporas. In C. Meillassoux (ed.) *The Development of Indigenous Trade and Markets in West Africa.* Oxford Unversity Press for the International African Institute, London. 285–302.

Critchfield, R. (1994) *The Villagers: Changed Values, Altered Lives: the Closing of the Urban-Rural Gap.* Anchor Books, New York.

Croll, Elisabeth (1983) *The Family Rice Bowl: Food and the Domestic Economy in China.* UN Research Institute for Social Development, Geneva, and Zed Press, London.

Davis J. M. and Svensgaard, D. J. (1987) Low level lead exposure and child development. *Nature* 329, 297–300.

Davis, Mike (1990) *City of Quartz: Excavating the Future in Los Angeles.* Verso, London.

Dayaratne, Ranjith and Samarawickrama, Raja (2001) Empowering Communities in Managing Rural Urban Encounters: the Concepts and Practices of the Housing and Community Development Program in the Peri-Urban areas of Colombo. Paper prepared for the conference: *Rural–urban Encounters: Managing the Environment of the Peri-urban Interface.* Development Planning Unit, University College London, 9–10 November.

de Haan, A. (1999) Livelihoods and poverty: the role of migration – a critical review of the migration literature. *Journal of Development Studies* 36 (2), 1–47.

de Haan, A. (2000) *Migrants, Livelihoods, and Rights: the Relevance of Migration in Development Policies.* Social Development Department, Department for International Development, Working Paper.

Development Planning Unit (no date) *Living between Urban and Rural Areas: Shaping Change for Improved Livelihoods and a Better Environment.* Volume 2*: Developing an Environmental Planning and Management Process for the Peri-urban Interface: Guiding and Working Principles.* Development Planning Unit, University College London.

Drakakis-Smith, David (1990) Food for thought or thought about food: urban food distribution systems in the Third World. In R. Potter and A. Salau (eds) *Cities and Development.* Mansell, London. 100–120.

Drakakis-Smith, David (1991) Urban food distribution in Asia and Africa. *The Geographical Journal* 157 (1), March, 51–61.

Drakakis-Smith, David (1992) Strategies for meeting basic food needs in Harare. In J. Baker and P. Ove Pedersen (eds) *The Rural–urban Interface in Africa.* Seminar Proceedings No. 27. Scandinavian Institute of African Studies, Uppsala.

Drakakis-Smith, D. (1996) Third World cities: sustainable urban development II – population, labour and poverty. *Urban Studies* 33 (4–5), 673–701.

Egziabher, A. G. (1994) Urban farming, co-operatives and the urban poor in Addis Ababa. In A. G. Egziabher *et al.* (eds) *Cities Feeding People: an Examination of Urban Agriculture in East Africa.* International Development Research Centre, Ottawa. 85–92.

Ellis, Frank (1983) Agricultural marketing and peasant–state transfers in Tanzania. *Journal of Peasant Studies* 10, 214–240.

Elson, Robert E. (1997) *The End of the Peasantry in Southeast Asia: a Social and Economic History of Peasant Livelihood, 1800–1990s*. Macmillan, Basingstoke.

Englund, Harri (2002) The village in the city, the city in the village; migrants and Lilongwe. *Journal of Southern African Studies* 28 (1), 137–154.

Epstein, T. S. (1982) *Urban Food Marketing and Third World Rural Development: the Structure of Producer-Seller Markets*. Croom Helm, London.

Epstein, T. S. (1973) *Rural India: Yesterday, Today, Tomorrow: Mysore villages*. Macmillan, London.

Fafchamps, M. and Minten, B. (1998) *Returns to Social Capital Among Traders*. Markets and Structural Studies Division Discussion Paper No. 23. International Food Policy Research Institute, Washington, DC. http://www.ifpri.org/divs/mtid/dp/mssdp23.htm [last accessed: 16 March 2004].

Fairhead, J. and Leach, M. (1992) *Misreading the African Landscape*. Cambridge University Press, Cambridge.

Fairhead, J. and Leach, M. (1995) Local agro-ecological management and forest–savanna transitions: the case of Kissidougou, Guinea. In T. Binns (ed.) *People and Environment in Africa*. John Wiley, Chichester. 163–170.

Fairhead, J. and Leach, M. (1998) *Reframing Deforestation: Global Analyses and Local Realities – Studies in West Africa*. Routledge, London.

Fang, S., Fang, C. and Zhang, X. (2003) Agricultural research and urban poverty: the case of China. *World Development* 31 (4), 733–741.

Feldman, Shelley (1999) Rural–urban linkages in South Asia: contemporary themes and policy directions. Paper presented at a workshop: *Poverty Reduction and Social Progress: New Trends and Emerging Lessons, a Regional Dialogue and Consultation on World Development Report 2001* for South Asia. Rajendrapu, Bangladesh, 4–6 April.

Findlay, A. M. and Findlay A. (1991) *Population and Development in the Third World*. Routledge, London.

Findlay, A. M., Paddison, Ronan and Dawson, John A. (eds) (1990) *Retailing Environments in Developing Countries*. Routledge, London.

Forbes, D. (1996) *Asian Metropolis: Urbanisation and the Southeast Asian City*. Oxford University Press, Melbourne.

Frank, A. G. (1967) *Capitalism and Underdevelopment in Latin America*. Monthly Review Press, New York.

Friedmann, J. (1966) *Regional Development Policy: a Case Study of Venezuela*. MIT Press, Cambridge, Mass.

Gilbert, A. and Gugler, J. (1992) *Cities, Poverty and Development in the Third World*. Wiley, Chichester.

Gittinger, J. P., Leslie, J. and Hoisington, C. (eds) (1987) *Food Policy: Integrating Supply, Distribution and Consumption*. Johns Hopkins University Press, Baltimore, Md.

Gould, W. T. S. (1969) The structure of space preferences in Tanzania. *Area* 1, 29–35.

Gugler, J. (1991) Life in a dual system revisited: urban–rural ties in Enugu, Nigeria, 1961–1987. *World Development* 19, 399–409.

Guyer, Jane (1987) Introduction. In Jane Guyer (ed.) *Feeding African Cities.* Manchester University Press, Manchester. 1–54.

Hagerstrand, T. (1952) *The Propagation of Innovation Waves*. Lund Studies in Geography, Lund.

Hansen, A. and McMillan, D. E. (eds) (1986) *Food in Sub-Saharan Africa.* Pinter, London.

Hardoy, J., Mitlin, D. and Satterthwaite, D. (2001) *Environmental Problems in an Urbanizing World: Finding Solutions for Cities in Africa, Asia and Latin America.* Earthscan, London.

Harriss, B. and Harriss, J. (1988) Generative or parasitic urbanism? Some observations from the recent history of a South Indian town market. *Journal of Development Studies* 20 (3), 82–101.

Harriss, J. and Moore, M. (eds) (1984) Development and the Rural–Urban Divide. Special issue of *The Journal of Development Studies* 20 (3).

Harvey, D. (1973) *Social Justice and the City*. Arnold, London.

Harvey, D. (1989) *The Condition of Postmodernity.* Blackwell, Oxford.

Heilig, G. (1999) *Can China Feed Itself? A System for Evaluation and Policy Analysis.* International Institute for Applied Systems Analysis, Vienna. http://www.iiasa.ac.at/Research/LUC/ChinaFood/Inde_m.htm [last accessed: 29 November 2003].

Hill, Polly (1963) Markets in Africa. *Journal of Modern African Studies* 1 (4), 441–453.

Hill, R. (1997) An agricultural transition on the Pacific Rim: explorations towards a model. In T. McGee and R. Waters (eds) *Asia Pacific: New Geographies of the Pacific Rim.* C. Hurst, New York. 93–112.

Hinderdink, J. and Titus, M. (1998) Paradigms of regional development and the role of small centres. In M. Titus and J. Hinderdink (eds) *Town and Hinterland in Developing Countries: Perspectives on Rural–Urban Interaction and Regional Development.* Thela Thesis, Amsterdam. 5–18.

Hirschman, A. (1958) *The Strategy of Economic Development.* Yale University Press, New Haven.

Hodder, R. (2000) *Development Geography.* Routledge, London.

Holm, M. (1994) Food supply and economic suitability in urban areas; a lesson from Tanzania. In M. A. Mohamed Saleh (ed.) *Inducing Food Security: Perspectives on Food Policies in Eastern and Southern Africa.* Seminar Procedings No. 30, Nordiska Afrikainstitutet, Uppsala.

Hulme, D. and Mosley, P. (1996a) *Finance against Poverty:* Volume 1. Routledge, London.

Hulme, D. and Mosley, P. (eds) (1996b) *Finance against Poverty,* Volume 2. Routledge, London.

Hulme, D., Montgomery, R. and Bhattarchaya, D. (1996) Mutual finance for the poor: a study of the Federation of Thrift and Credit Cooperatives (SANASA) in Sri Lanka. In D. Hulme and P. Mosley (eds) *Finance against Poverty,* Volume 2. Routledge, London. 177–245.

Iaquinta, David and Drescher, Axel (2001) More than the spatial fringe: an application of the periurban typology to planning and management of natural resources. Paper presented to the conference: *Rural–Urban Encounters: Managing the Environment of the Peri-Urban Interface*. Development Planning Unit, University College London, 9–10 November.

Iliya, M. (1999) *Income Diversification in the Semi-arid Zone of Nigeria: a Study of Gigane, Sokoto, North-west Nigeria.* De-Agrarianisation and Rural Employment (DARE) Network. African Studies Centre Working Paper 39. University of Leiden.

Independent Commission on International Development Issues (1980) *North–South: a Programme for Survival: Report of the Independent Commission.* Pan, London. (Also known as the 'Brandt Report'.)

ICPD (International Conference on Population and Development) (1994) *Programme of Action Summary.* United Nations Population Fund. http://www.unfpa.org/icpd/summary.htm [last accessed: 5 March 2004).

International Telecommunications Union (1997) *Report on Communications for Rural and Remote Areas*. Telecommunication Development Bureau, ITU, Geneva. Document 2/224. http://www.itu.int/ITU-D/univ_access/reports/mso2bd.pdf [last accessed: 27 September 2004].

International Telecommunications Union (2001) *African Telecommunications Indicators 2001.* International Telecommunications Union, Geneva.

Jamal, V. and Weeks, J. (1988) The vanishing rural–urban gap in sub-Saharan Africa. *International Labour Review* 127 (3), 271–292.

Jambiya, G. (1998) *The Dynamics of Population, Land Scarcity, Agriculture and Non-agricultural Activities: West Usambara Mountains, Lushoto District, Tanzania.* De-Agrarianisation and Rural Employment (DARE) Network. African Studies Centre Working Paper 28. University of Leiden.

Johnson, S. (1995) *The Politics of Population: the International Conference on Population and Development, Cairo, 1994.* Earthscan, London.

Jones, William O. (1972) *Marketing Staple Food Crops in Tropical Africa.* Cornell University Press, Ithaca, NY.

Katz, C. (1997) Gender and trade within the household. observations from rural Guatemala. *World Development* 23 (2), 327–342.

Kaynak, E. (1981) Food distribution systems: evolution in Latin America and the Middle East. *Food Policy*, May, 78–90.

Kaynak, E. (1986) *Marketing and Economic Development.* Praeger Publishers, Newark, NJ.

Kleih, U., Odwongo, W. and Ndyashangaki, C. (1999) *Community Access to Marketing Opportunities: Options for Remote Areas; Uganda Case Study.* Natural Resources Institute, Greenwich, and Agricultural Policy Secretariat, Kampala.

Knight, J. and Song, L. (2000) *The Rural–Urban Divide: Economic Disparities and Interactions in China*. Oxford University Press, Oxford.

Koc, M., MacRae, R., Mougeot, L. J. A. and Walsh J. (eds) (1999) *Hunger-Proof Cities: Sustainable Urban Food Systems.* International Development Research Centre, Ottawa.

Kynge, J. (2002) Taxman heaps added burden on farmers' backs. *Financial Times*, 31 October, 13.

Lado, C. (1987) Some aspects of rural marketing systems and peasant farming in Maridi District, Southern Sudan. *Transactions of the Institute of British Geographers* 13, 361–374.

Leach, M. and Mearns, R. (1988) *Beyond the Woodfuel Crisis: People, Land and Trees in Africa.* Earthscan, London.

Lee-Smith, D. and Memon, P. A. (1994) Urban Agriculture in Kenya. In A. G. Egziabher *et al.* (eds) *Cities Feeding People: an Examination of Urban Agriculture in East Africa.* International Development Research Centre, Ottawa. 67–84.

Lester, A. (1998) *From Colonization to Democracy: a New Historical Geography of South Africa.* I. B. Taurus, London.

Lester, A., Nel, E. and Binns, T. (2000) *South Africa; Past, Present and Future – Gold at the End of the Rainbow?* Prentice Hall, Harlow.

Lewcock, C. (1995) Farmer use of urban waste in Kano. *Habitat International* 19 (2), 225–234.

Liebes, T. and Katz, E. (1989) On the critical abilities of television viewers. In E. Seiter, H. Borchers, G. Kreutzner and E. M. Warth (eds) *Remote Control: Television, Audiences and Cultural Power*. Routledge, London. 204–222.

Lipton, M. (1977) *Why Poor People Stay Poor: a Study of Urban Bias in World Development.* Temple Smith, London.

Lipton, M. (1984) Urban bias revisited. *Journal of Development Studies* 20 (3), 139–166.

Lipton, M. (1993) Urban bias – of consequences, classes and causality. *Journal of Development Studies* 29 (4), 229–258.

Lynch, K. (1992) *The production, distribution and marketing of fruit and vegetables for the urban market of Dar es Salaam, Tanzania.* Unpublished PhD thesis, University of Glasgow.

Lynch, K. (1995) Sustainability and urban food supply in Africa. *Sustainable Development* 3, 79–88.

Lynch, K. (2002) Urban agriculture. In V. Desai and R. Potter (eds) *Arnold Companion to Development Studies.* Edward Arnold, London. 268–272.

Lynch, K. (2004) Managing urbanisation. In F. Harris (ed.) *Global Environmental Issues*. John Wiley, Chichester. 195–228.

Lynch, K. D., Binns, T. and Olofin, E. A. (2001) Urban agriculture under threat – the land security question in Kano, Nigeria. *Cities* 18 (3), 159–171.

Lynch, K. D., Poole, N. and Ashimogo, G. (2001) *ICT technology for overcoming market information constraints in East Africa.* Unpublished paper prepared for the Crop Post Harvest Programme, Department for International Development. Kingston University, London.

Macaulay, J. (2003) Timor Leste: newest and poorest of Asian nations. *Geography* 88 (1), 40–46.

McGee, T. G. (1971) *The Urbanisation Process in the Third World: Explorations in Search of a Theory*. G. Bell and Sons, London.

McGee, T. G. (1991) The emergence of desakota regions in Asia: expanding a hypothesis. In N. Ginsberg, B. Poppel and T. G. McGee (eds) *The Extended Metropolis*. University of Hawaii Press, Honolulu. 3–25.

McGranahan, G. and Satterthwaite, D. (2000) Environmental health or ecological sustainability? Reconciling the brown and green agendas in urban development. In Cedric Pugh (ed.) *Sustainable Cities in Developing Countries*. Earthscan, London. 73–90.

McGranahan, G. and Satterthwaite, D. (2002) The environmental dimensions of sustainable development for cities. *Geography* 87 (3), 213–226.

Madulu, N. F. (1998) *Changing Lifestyles in Farming Societies of Sukumaland: Kwimba District, Tanzania*. De-Agrarianisation and Rural Employment (DARE) Network. African Studies Centre, Working Paper 27. University of Leiden.

Main, H. (1995) The effects of urbanisation in rural environments in Africa. In T. Binns (ed.) *People and Environment in Africa*. John Wiley, Chichester. 47–57.

Maluf, R. (1998) Economic development and the food question in Latin America. *Food Policy* 23 (2), 155–172.

Mandanipour, A. (1998) *Tehran: the Making of a Metropolis*. John Wiley, Chichester.

Mankekar, Purnima (2000) National texts and gendered lives: an ethnography of television viewers in a north Indian city. In Kelly Askew and Richard Wilk (eds) *The Anthropology of Media: a Reader*. Blackwells, Oxford. 299–322. (Reprinted from *American Ethnologist* 20 (3), 543–563.)

Mbiba, B. (2001) Peri-Urban transformations and livelihoods in East and Southern Africa: insights from the peri-net research experience. Paper presented at the conference: *Rural–Urban Encounters: Managing the Environment of the Peri-Urban Interface*. Development Planning Institute, University College London, 9–10 November.

Meagher, K. (1997) Shifting the imbalance – the impact of structural adjustment on rural–urban population movements in northern Nigeria. *Journal of Asian and African Studies* 32 (1–2), 81–92.

Meagher, K. (1999) *If the Drumming Changes, the Dance Also Changes: De-agrarianisation and Rural Non-farm Employment in the Nigerian Savannah*. De-Agrarianisation and Rural Employment (DARE) Network. African Studies Centre Working Paper 40. University of Leiden.

Mearns, R. (1995) Institutions and natural resource management: access to and control over woodfuel in East Africa. In T. Binns (ed.) *People and Environment in Africa*. John Wiley, Chichester. 103–114.

Mehmet, O. (1995) *Westernising the Third World: the Euro-Centricity of Economic Development Theories*. Routledge, London.

Mincer, J. (1978) Family migration decisions. *Journal of Political Economy* 86 (5), 749–773.

Montgomery, R., Battarchaya, D. and Hulme, D. (1996) Credit for the poor in Bangladesh; the BRAC rural development programme and the government Thana resource development and employment programme. In D. Hulme and P. Mosley (eds) *Finance against Poverty,* Volume 2. Routledge, London. 94–176.

Morris, A. (1998a) *Geography and Development.* UCL Press, London.

Morris, A. (1998b) Market behaviour and market systems in the state of Mexico. In P. van Limbert and O. Verkoren (eds) *Small Towns and Beyond: Rural Transformation and Small Urban Centres in Latin America.* Thela Publishers, Amsterdam.

Mortimore, M. (1989) *Adapting to Drought; Farmers, Famines and Desertification in West Africa.* Cambridge University Press, Cambridge.

Mortimore, M. (1998) *Roots in the African Dust; Sustaining the Drylands.* Cambridge University Press, Cambridge.

Mortimore, M. and Tiffen, M. (1995) Population and environment in time perspective: the Machakos story. In T. Binns (ed.) *People and Environment in Africa.* John Wiley, Chichester. 69–89.

Mosley, P. (1983) *The Settler Economies: Studies in the Economic History of Kenya and Southern Rhodesia 1900–1963.* Cambridge University Press, Cambridge.

Mosley, P. (1996) Metamorphosis from NGO to commercial bank: the case of Bancosol in Bolivia. In D. Hulme and P. Mosley (eds) *Finance against Poverty,* Volume 2. Routledge, London. 25–58.

Mosse, D., Gupta, S., Mehta, M., Shah, V., Rees, J. and the KRIBP Project Team (2002) Brokered livelihoods: debt, labour migration and development in tribal western India. *Journal of Development Studies.* Special issue Labour Mobility and Rural Society. 38 (5), 59–88.

Mougeot, L. J. A. (1999) *Urban Agriculture: Definition, Presence, Potential and Risks, Main Policy Challenges.* Cities Feeding People, CFP Report Series. Report 31. International Development Research Centre, Ottawa. http://network.idrc.ca/ev.php?ID=2571_201&ID2=DO_TOPIC [last accessed: 16 March 2004].

Murray, C. (2002) Rural livelihoods. In V. Desai and R. Potter (eds) *Arnold Companion to Development Studies.* Edward Arnold, London, 151–155.

Mustapha, R. (1999) *Cocoa Farming and Income Diversification in Southwestern Nigeria.* De-Agrarianisation and Rural Employment (DARE) Network. African Studies Centre Working Paper 42. University of Leiden.

Mutizwa-Mangiza, Naison (1999) Strengthening rural–urban linkages. *Habitat Debate* 5 (1), 1–6.

Mutula, Stephen M. (2001) Internet access in East Africa: a future outlook. *Library Review* 50 (1), 28–33.

Mvena, Z. S. K., Lupanga, I. J. and Mlosi, M. R. S. (1991) *Urban Agriculture in Tanzania: a Study of Six Towns.* Sokoine University of Agriculture, Morogoro, and International Development Research Centre, Ontario.

Myint, H. (1964) *The Economics of Developing Countries.* Hutchinson, London.

Nunan, Fiona, Bird, Kate and Bishop, Joshua, with Anthony Edmundson and S. R. Nidagundi (no date) *Valuing Peri-urban Natural Resources: a Guide for Natural Resources Managers*. School of Public Policy, University of Birmingham.

O'Connor, A. (1983) *The African City*. Hutchinson University Library for Africa, London.

ODI (2002) *Rural–urban linkages*. Key Sheets for Sustainable Livelihoods 10. Overseas Development Institute, London. http://www.odi.org.uk/keysheets/ [last accessed: 28 November 2003].

OECD (1996) *Shaping the Urban Environment in the 21st Century*. Organisation for Economic Cooperation and Development, Development Assistance Committee, Paris. http://www.oecd.org/dac/urbenv/index.htm [last accessed: 1 December 2003].

O'Farrell, C., Norrish, P. and Scott, A. (2000) Information and Communication Technologies for Sustainable Livelihoods. *SD Dimensions*. http://www.fao.org/sd/cddirect/cdre0055d.htm [last accessed: 12 March 2004]

O'Meara, Molly (1999) *Reinventing Cities for People and the Planet*. WorldWatch Paper 147. http://secure.worldwatch.org/cgi-bin/wwinst/BWP147 [last accessed: 28 November 2003].

Pacione, M. (2001) *Urban Geography; a Global Perspective*. Routledge, London.

Paddison, R., Findlay, A. and Dawson, A. (1990) Retailing in less-developed countries. In A. M. Findlay, R. Paddison and J. A. Dawson (eds) *Retailing Environments in Developing Countries*. Routledge, London.

Painter, Thomas M. (1996) Space, time and rural urban linkages in Africa: notes for a geography of livelihoods. *African Rural and Urban Studies* 3 (1), 79–98.

Parnwell, M. (1993) *Population Movements and the Third World*. Routledge, London.

Payne, G. (2002) Tenure and shelter in urban livelihoods. In C. Rakodi with T. Lloyd-Jones (eds) *Urban Livelihoods: a People-Centred Approach to Reducing Poverty*. Earthscan, London. 151–164.

Piault, M. (1971) Cycles de marchés et 'espaces' socio-politique. In C. Meillassoux (ed). *The Development of Trade and Markets in West Africa*. Oxford University Press for the International African Institute, London. 285–302.

Ping, H., Fang, C., Biao, X. and Chunhua, M. (1999) Rural migration and rural development: a report on the field investigation of eight villages from four provinces in China. In S. Fitzpatrick (ed.) *Work and Mobility: Recent Labour Migration Issues in China*. Working Paper 6. Asia Pacific Migration Research Network (APMRN), Woolongong.

Pletcher, J. R. (1982) *Agricultural policy and the political crisis in Zambia*. Ph.D. thesis, University of Wisconsin.

Poole, N. D. and Lynch, K. D. (2003) Agricultural market knowledge appropriate delivery systems for a private and public good? *Journal of Agricultural Education and Extension* 9 (3).

Poole, N. D., Marshall, F. and Bhupal, D. S. (2002) Air pollution effects and initiatives to improve food quality assurance in India. *Quarterly Journal of International Agriculture* 41(4), 363–385.

Potter, R. and Lloyd-Evans, S. (1998) *The City in the Developing World.* Longman, Harlow.

Potter, R., Binns, T., Elliott, J. and Smith, D. (1999) *Geographies of Development.* Longman, Harlow.

Potter, R., Binns, T., Elliott, J. and Smith, D. (2004) *Geographies of Development.* Pearson, Harlow. Second edition.

Potts, D. (1995) Shall we go home? Increasing urban poverty in African cities. *Geographical Journal* 161 (3), 245–264.

Potts, D. (1997) Urban lives: adopting new strategies and adapting rural links. In Carole Rakodi (ed.) *The Urban Challenge in Africa: Growth and Management of Its Large Cities.* United Nations University, Tokyo. 447–494.

Potts, D. with Mutambirwa, C. (1998) Basics are now a luxury: perceptions of ESAP's impact on rural and urban areas in Zimbabwe. *Environment and Urbanization* 10 (1), 55–75.

Pugh, C. (1996) 'Urban bias', the political economy of development and urban policies for developing countries. *Urban Studies* 33 (7), 1045–1060.

Qadeer, M. A. (2000) Ruralopolises: the spatial organisations of residential land economy of high density rural regions in South Asia. *Urban Studies* 37 (9), 1583–1603.

Rakodi, C. (1988) Self-reliance or survival? Food production in African cities with particular reference to Zambia. Paper presented to the conference: Urban Food supplies and Peri-Urban Agriculture Workshop. Centre for African Studies, School of Oriental and African Studies, 6 May 1988.

Rakodi, C. (2002) Economic development, urbanization and poverty. In C. Rakodi with T. Lloyd-Jones (eds) *Urban Livelihoods; a People-Centred Approach to Reducing Poverty.* Earthscan, London. 23–36.

Ramaswami, B. and Balakrishnan, P. (2002) Food prices and the efficiency of public intervention: the case of the public distribution system in India. *Food Policy* 217, 419–436.

Reardon, T., Berdegué, J. and Escobar, G. (2001) Rural nonfarm employment and incomes in Latin America: overview and policy implications. *World Development* 29 (3), 395–409.

Rees, W. (1992) Ecological footprints and carrying capacity: what urban economics leaves out. *Environment and Urbanization* 4 (2), 121–130.

Rees, W. (1997) Is 'sustainable city' an oxymoron? *Local Environment* 2 (3), 303–310.

Regmi, C. and Tisdell, T. (2002) Remitting behaviour of Nepalese rural-to-urban migrants: implications for theory and policy. *Journal of Development Studies* 38 (3), 76–94.

Richardson, D., Ramirez, R. and Haq, M. (2000) *Grameen Telecom's Village Phone Programme: a Multi-Media Case Study.* TeleCommons Development

Group, Guelph. http://www.telecommons.com/villagephone/index.html [last accessed: 12 March 2004].

Riddell, B. (1997) Structural adjustment and the city in tropical Africa. *Urban Studies* 34 (8), 1297–1307.

Rider Smith, D. (1999) *Linking Rural and Urban Strategies: Department for International Development Portfolio Review*. Social and Economic Development Series. Natural Resources Institute, Chatham, UK.

Rigg, J. (1991) *South East Asia, a Region in Transition: a Thematic Human Geography of the ASEAN Region*. Unwin Hyman, London.

Rigg, J. (1998a) Tracking the poor: the making of wealth and poverty in Thailand (1982–1994). *Journal of Social Economics* 25 (6/7/8), 1128–1141.

Rigg, J. (1998b) Rural–urban interactions, agriculture and wealth: a southeast Asian perspective. *Progress in Human Geography* 22 (4), 497–522.

Roberts, B. R. (1995) *The Making of Citizens: Cities of Peasants Revisited*. Arnold, London.

Robinson, G. (2004) *Geographies of Agriculture: Globalisation, Restructuring and Sustainability*. Pearson, Harlow.

Robinson, M. (2001) *The Microfinance Revolution: Sustainable Finances for the Poor*. World Bank, Washington, and the Open Society Institute, New York.

Rogerson, C. (2001) In search of the African miracle: successful small enterprise development in Africa. *Habitat International* 25, 115–142.

Rondinelli, D. (1983) *Secondary Cities in Developing Countries. Policies for Diffusing Urbanisation*. Sage, Beverly Hills, Calif.

Rondinelli, D. and Ruddle, K. (1978) *Urbanisation and Rural Development: a Spatial Policy for Equitable Growth*. Praeger, New York.

Rostow, W. W. (1960) *The Stages of Economic Growth: a Non-communist Manifesto*. Cambridge University Press, Cambridge.

Rostow, W. W. (1990) *The Stages of Economic Growth: a Non-communist Manifesto*. Cambridge University Press, Cambridge. Second edition.

Rotgé, V. L. (2000) *Rural–Urban Integration in Java: Consequences for Regional Development and Employment*. Ashgate Publishing, Aldershot.

Rutherford, S., Harper, M. and Grierson, J. (2002) Support for livelihood strategies. In C. Rakodi (ed.) *Urban Livelihoods; a People-Centred Approach to Reducing Poverty*. Earthscan, London. 112–132.

Samiee, Saeed (1990) Impediments to retailing in developing countries. In A. Findlay, R. Paddison and J. Dawson (eds) *Retailing Environments in Developing Countries*. Routledge, London. 30–39.

Satterthwaite, D. (2000) Seeking an understanding of poverty that recognizes rural–urban differences and rural–urban linkages. Paper presented at the World Bank's *Urban Forum on Urban Poverty Reduction in the 21st Century*, April 2000.

Sawio, C. (1994) Who are the farmers of Dar es Salaam? In A. G. Agziabher, D. Lee-Smith, D. G. Maxwell, P. A. Memon, L. J. A. Mougeot and C. Sawio (eds) *Cities Feeding People: an Examination of Urban Agriculture in East Africa*. International Development Research Centre, Ottawa. 25–45.

Schell, L. M., Smith, M.T. and Bilsborough, A. (eds) (1993) *Urban Ecology and Health in the Third World.* Cambridge University Press, Cambridge.

Schilderman, Theo (2002) *Strengthening the Knowledge and Information Systems of the Urban Poor.* Intermediate Technology Development Group and Department for International Development, London. http://www.itdg.org/html/shelter/docs/kis_urban_poor_report_march2002.doc [last accessed: 1 December 2003]

Schneider, H. (1999) Livelihood strategies of urban households in secondary cities in Thailand and the Philippines – a comparison. In G. P. Chapman, A. K. Dutt and R. W. Bradnock (eds) *Urban Growth and Development in Asia,* Volume II*: Living in the Cities.* SOAS Studies in Development Geography. Ashgate, Aldershot. 222–243.

Sen, Amartya (1982) *Poverty and Famine.* Clarendon Press, Oxford.

Shepherd, A. W. (1997) Market information services. *AGS Bulletin.* FAO, Rome.

Shettima, K. A. (1997) Ecology, identity, developmentalism and displacement in northern Nigeria. *Journal of Asian and African Studies* 32 (1–2), 66–80.

Simon, D., Nsiah-Gyabaah, K., Warburton , H., Adu-Gyamfi, V. and McGregor, D. (2001) *The Changing Urban–Rural Interface of African Cities: Conceptual Issues and an Application to Kumasi, Ghana.* CEDAR Kumasi Paper 1, Ghana.

Simpson, E. S. (1996) *The Developing World: an Introduction*, Longman, Harlow.

Singh, J. P. (1986) *Patterns of Rural–Urban Migration in India.* Inter-India Publications, New Delhi.

Skinner, G. W. (1964/65) Marketing and social structure in rural China, Parts I and II. *Journal of Asian Studies* 24, 3–43 and 195–228.

Slater, D. (1974) Colonialism and the spatial structure of underdevelopment: outlines of an alternative approach, with special reference to Tanzania. *Progress in Planning* 4, Part V, 146–159.

Smit, Warren (1998) The rural linkages of urban households in Durban, South Africa. *Environment and Urbanization* 10 (1), 77–87.

Smith, D. M. (ed.) (1992) *The Apartheid City and Beyond: Urbanization and Social Change in South Africa.* Routledge, London.

Smith, D. M. (1994) *Geography and Social Justice.* Blackwells, London.

Smith, D. W. (1998) Urban food systems and the poor in developing countries. *Transactions of the Institute of British Geography* 23 (2), 207–220.

Snrech, S. (1996) *Etats des réflexions sur les transformations de l'agriculture dans le Sahel.* Paper No. SAH/D(96)451. Club du Sahel/OCDE, Paris.

Soja, E. W. (1989) *Postmodern Geographics: the Reassertion of Space in Critical Social Theory.* Verso, London.

Soussan, J., O'Keefe, P. and Munslow, B. (1990) Urban fuelwood – challenges and dilemmas. *Energy Policy* 18 (6), 572–582.

Sporrek, A. (1985) *Food Marketing and Urban Growth in Dar es Salaam.* Department of Geography, Royal University of Lund.

Sreberny-Mohammadi, A. (2002) The global and the local in international communications. In Kelly Askew and Richard Wilk (eds) *The Anthropology of Media: a Reader*. Blackwells, Oxford. 337–356. (Reprinted from James Curran and Michael Guerevitch (eds) (1991) *Mass Media and Society*. Edward Arnold, London. 118–138.)

Stephens, C. (2000) Inequalities in urban environments, health and power: reflections on theory and practice. In C. Pugh (ed.) *Sustainable Cities in Developing Countries*. Earthscan, London. 91–114.

Stevens, L. and Rule, S. (1999) Moving to an informal settlement: the Gauteng experience. *The South African Geographical Journal* 81 (3), April. http://www.egs.uct.ac.za/sagj/Gauteng.htm [last accessed: 1 December 2003]

Swaminathan, M. (1999) Understanding the costs of the food corporation of India. *Economic and Political Weekly* 34 (52), A121–A130.

Tacoli, C. (1998a) *Bridging the Divide: Rural–Urban Interaction and Livelihood Strategies*. Gatekeeper Series No. 11. International Institute for Environment and Development, London.

Tacoli, Cecilia (1998b) Beyond the rural–urban divide. Special edition of *Environment and Urbanization* 10 (1).

Tacoli, Cecilia (1999) Understanding the opportunities and constraints for low-income groups in the peri-urban interface: the contribution of livelihood frameworks. *Strategic Environmental Planning and Management for the Peri-urban Interface*. Research Project Discussion Paper. http://www.ucl.ac.uk/dpu/pui [last accessed: 30 November 2003].

Tiffen, M. (2003) Transition in sub-Saharan Africa: agriculture, urbanization and income growth. *World Development* 31 (8), 1343–1366.

Tiffen, M., Mortimore, M. and Gichuki, F. (1994) *More People, Less Erosion. Environmental Recovery in Kenya*. John Wiley, Chichester.

Todaro, M. P. (2000) *Economic Development*. Addison Wesley, Harlow. Seventh edition.

Toye, J. (1991) Ghana. In P. Mosley, J. Harrington and J. Toye (eds) *Aid and Power: the World Bank and Policy-Based Lending,* Volume 2: *Case Studies*. Routledge, London. 151–199.

Trager, Lillian (1996) Mobility, linkages and 'local' institutions in African development. *African Rural and Urban Studies* 3 (1), 7–24.

UNCHS (United Nations Commission for Human Settlements) (1996) *Habitat II Conference: Habitat Agenda*. United Nations Commission for Human Settlements, Nairobi. http://www.unhabitat.org/unchs/english/hagenda/ [last accessed: 16 March 2004].

UNCHS (United Nations Commission for Human Settlements) (1999) *World Urbanisation Prospects 1999 Revision*. United Nations Commission for Human Settlements, Nairobi.

UNCHS (United Nations Commission for Human Settlements) (2001) *Cities in a Globalizing World. Global Report on Human Settlements 2001*. United Nations Commission for Human Settlements, New York. http://www.earthscan.co.uk/cities/contents.htm [last accessed: 17 March 2004].

United Nations (2003) *World Urbanization Prospects*. United Nations, New York. http://www.un.org/esa/population/publications/wup2003/2003wup.htm [last accessed: 17 May 2004].

UNDP/UNCHS (Habitat) (1995) Rural–urban linkages: policy guidelines for rural development. Paper presented for the Twenty-Third Meeting of the ACC Sub-Committee on Rural Development. UNESCO, Paris, 31 May–2 June.

UNESCAP (United Nations Economic and Social Commission for Asia and the Pacific) (2001) *Reducing Disparities: Balanced Development of Urban and Rural Areas and Regions within the Countries of Asia and the Pacific*. United Nations, New York, 2001. http://www.unescap.org/huset/disparities/ [last accessed: 5 March 2004].

UNFPA (1996) Urban-Rural Links: Transactions and Transformations. *State of World Population*. United Nations Population Fund, New York. http://www.unfpa.org/swp/1996/SWP96MN.HTM [last accessed: 28 November 2003].

UNFPA (2003) *State of World Population*. United Nations Population Fund http://www.unfpa.org/swp/2003/ [last accessed: 28 November 03].

Urban Agriculture Network (1996) *Urban Agriculture; Food, Jobs and Sustainable Cities*. United Nations Development Programme, New York.

van Donge, J. K. (1992a) Waluguru traders in Dar es Salaam: an analysis of the social construction of economic life. *African Affairs* 91 (363), 181–205.

van Donge, J. K. (1992b) Agricultural decline in Tanzania: the case of the Uluguru Mountains. *African Affairs* 91 (362), 73–94.

Vance, J. (1970) *The Merchant's World: the Geography of Wholesaling*. Prentice Hall, Englewood Cliffs, NJ.

von Braun, J. and Kennedy E. (eds) (1994) *Commercialisation of Agriculture, Economic Development and Nutrition*. Johns Hopkins University Press, Baltimore, Md.

von Braun, J., McComb, J., Mensah, B. K. and Pandya-Lorch, R. (1993) *Urban Food Insecurity and Malnutrition in Developing Countries; Trends, Policies and Research Implications*. International Food Policy Research Institute, Washington.

Wackernagel, M. (1998) The ecological footprint of Santiago de Chile. *Local Environment* 3 (1), 7–25.

Wallis, W. (1999a) Cocoa: a hot political issue. *Financial Times*. Ghana Country Survey. 4 November, p. VII.

Wallis, W. (1999b) Tasty produce, sour prices. *Financial Times*. Ghana Country Survey. 4 November, p. VII.

Wallis, W. (2000) Cocoa: falling price hits economy's backbone. *Financial Times*. Ghana Country Survey. 4 November, p. VI.

Wilhelm, L. (1997) *Food Supply and Distribution Networks and the Functioning of Markets in Africa*. Food Into Cities Collection. Food and Agriculture Organisation, Rome.

Wilk, Richard (2002) 'It's destroying a whole generation': television and moral discourse in Belize. In Kelly Askew and Richard Wilk (eds) *The Anthropology of Media: a Reader*. Blackwells, Oxford. 286–298. (Reprinted from *Visual Anthropology* 5 (3/4), 1993, 229–244.)

World Bank (1999) *World Development Report 1999*. Oxford University Press, Oxford. http://www.worldbank.org/wdr/WDR1999/ [last accessed: 17 March 2004].

World Bank (2000) *Rural–Urban Linkages and Interactions: Synthesis of Issues, Conclusions and Priority Opportunities Emerging from the 9 March Workshop*. World Bank, New York.

World Bank (2002) *World Development Report 2002*. World Bank, New York. http://www.worldbank.org/wdr/WDR2002/ [last accessed: 17 March 2004].

World Bank (2003) *Bangladesh: Data Profile*. World Bank, New York. http://www.worldbank.org/data/dataquery.html [last accessed: 2 December 2003].

World Commission on Environment and Development (WCED) (1987) *Our Common Future*. Oxford University Press, Oxford. (Also known as the 'Brundtland Report'.)

Youssry, Mahmoud and Aboul Atta, Tarek A. (1997) The challenge of urban growth in Cairo. In C. Rakodi (ed.) *The Urban Challenge in Africa; Growth and Management of its Large Cities*. United Nations University Press, Tokyo. 111–149.

Yunusa, M. B. (1999) *Not Farms Alone: a Study of Rural Livelihoods in the Middle Belt of Nigeria*. De-Agrarianisation and Rural Employment (DARE) Network, African Studies Centre Working Paper 38. University of Leiden.

Zhu, Y. (2000) In situ urbanization in rural China: case studies from Fujian Province. *Development And Change* 31 (2), 413–434.

Zongo, G. (2001) *Information and Communication Technologies for Development in Africa: Trends and Overview*. International Development Research Centre/Acacia Communities and Information Society in Africa, Dakar.

Index